Studies in Classification, Data Analysis, and Knowledge Organization

Springer
Berlin
Heidelberg
New York
Hong Kong
London
Milan
Paris
Tokyo

Titles in the Series

H.-H. Bock and P. Ihm (Eds.)
Classification, Data Analysis, and Knowledge Organization. 1991 (out of print)

M. Schader (Ed.)
Analyzing and Modeling Data and Knowledge. 1992

O. Opitz, B. Lausen, and R. Klar (Eds.)
Information and Classification. 1993 (out of print)

H.-H. Bock, W. Lenski, and M.M. Richter (Eds.)
Information Systems and Data Analysis. 1994 (out of print)

E. Diday, Y. Lechevallier, M. Schader, P. Bertrand, and B. Burtschy (Eds.)
New Approaches in Classification and Data Analysis. 1994 (out of print)

W. Gaul and D. Pfeifer (Eds.)
From Data to Knowledge. 1995

H.-H. Bock and W. Polasek (Eds.)
Data Analysis and Information Systems. 1996

E. Diday, Y. Lechevallier and O. Opitz (Eds.)
Ordinal and Symbolic Data Analysis. 1996

R. Klar and O. Opitz (Eds.)
Classification and Knowledge Organization. 1997

C. Hayashi, N. Ohsumi, K. Yajima, Y. Tanaka, H.-H. Bock, and Y. Baba (Eds.)
Data Science, Classification, and Related Methods. 1998

I. Balderjahn, R. Mathar, and M. Schader (Eds.)
Classification, Data Analysis, and Data Highways. 1998

A. Rizzi, M. Vichi, and H.-H. Bock (Eds.)
Advances in Data Science and Classification. 1998

M. Vichi and O. Opitz (Eds.)
Classification and Data Analysis. 1999

W. Gaul and H. Locarek-Junge (Eds.)
Classification in the Information Age. 1999

H.-H. Bock and E. Diday (Eds.)
Analysis of Symbolic Data. 2000

H. A. L. Kiers, J.-P. Rasson, P. J. F. Groenen, and M. Schader (Eds.)
Data Analysis, Classification, and Related Methods. 2000

W. Gaul, O. Opitz and M. Schader (Eds.)
Data Analysis. 2000

R. Decker and W. Gaul
Classification and Information Processing at the Turn of the Millenium. 2000

S. Borra, R. Rocci, M. Vichi, and M. Schader (Eds.)
Advances in Classification and Data Analysis. 2001

W. Gaul and G. Ritter (Eds.)
Classification, Automation, and New Media. 2002

K. Jajuga, A. Sokołowski, and H.-H. Bock (Eds.)
Classification, Clustering and Data Analysis. 2002

M. Schwaiger, O. Opitz (Eds.)
Exploratory Data Analysis in Empirical Research. 2003

M. Schader, W. Gaul, and M. Vichi (Eds.)
Between Data Science and Applied Data Analysis. 2003

Hans-Hermann Bock Marcello Chiodi
Antonino Mineo
Editors

Advances in Multivariate Data Analysis

Proceedings of the Meeting of the Classification and Data Analysis Group (CLADAG) of the Italian Statistical Society, University of Palermo, July 5–6, 2001

With 48 Figures and 56 Tables

Springer

Prof. Dr. Hans-Hermann Bock
Institute of Statistics
RWTH Aachen University
Wuellnerstr. 3
52056 Aachen
Germany
bock@stochastik.rwth-aachen.de

Prof. Marcello Chiodi
Prof. Antonino Mineo

Department of Statistics and Mathematics Silvio Vianelli
University of Palermo
Viale delle Scienze
90128 Palermo
Italy
chiodi@www.unipa.it
amineo@www.unipa.it

ISBN 978-3-540-20889-1 ISBN 978-3-642-17111-6 (eBook)
DOI 10.1007/978-3-642-17111-6

Library of Congress Control Number: 2004102976

springeronline.com

Originally published by Springer-Verlag Berlin Heidelberg New York in 2004

Softcover-Design: Erich Kirchner, Heidelberg

SPIN 10984314 43/3130/DK – 5 4 3 2 1 0 – Printed on acid-free paper

Preface

This volume contains a selection of papers presented during the biennial meeting of the CLAssification and Data Analysis Group (CLADAG) of the Società Italiana di Statistica which was organized by the Istituto di Statistica of the Università degli Studi di Palermo and held in the Palazzo Steri in Palermo on July 5-6, 2001. For this conference, and after checking the submitted 4-page abstracts, 54 papers were admitted for presentation. They covered a large range of topics from multivariate data analysis, with special emphasis on classification and clustering, computational statistics, time series analysis, and applications in various classical or recent domains. A two-fold careful reviewing process led to the selection of 22 papers which are presented in this volume. They convey either a new idea or methodology, present a new algorithm, or concern an interesting application.

We have clustered these papers into five groups as follows:

1. Classification Methods with Applications
2. Time Series Analysis and Related Methods
3. Computer Intensive Techniques and Algorithms
4. Classification and Data Analysis in Economics
5. Multivariate Analysis in Applied Sciences.

In each section the papers are arranged in alphabetical order.

The editors - two of them the organizers of the CLADAG conference - would like to express their gratitude to the authors whose enthusiastic participation made the meeting possible and very successful. We also want to extend our thanks to the international group of reviewers for their diligence and the time spent in their professional refereeing work. Moreover, we are grateful to the chairpersons and discussants of the 13 sessions of the conference, their comments provided useful suggestions for the authors and the audience.

Special thanks are due to the Local Organizing Committee from the University of Palermo that comprised

Salvatore Bologna, Angelo M. Mineo, Antonella Plaia

together with Antonino Mineo (coordinator) and Marcello Chiodi who have co-edited this volume.

Finally, we would like to thank Dr. Martina Bihn from Springer Verlag for the excellent cooperation in publishing this volume.

Palermo and Aachen

Hans-Hermann Bock
Marcello Chiodi
Antonino Mineo

List of Referees

We thank the following scientists for their expertness and diligence in reviewing the papers of this volume:

Pietro Amenta, Lecce
Thomas Bausch, Garmish-Partenkirchen
Andrea Cerioli, Parma
Cathrine Charles, Namur
Juergen Hansohm, Munchen
Roger Lafosse, Toulouse
Berthold Lausen, Erlangen-Nürnberg
Yves Lechevallier, Rocquencourt
Herbert Lee, Santa Cruz
Gianfranco Lovison, Palermo
Magdalena Mißler-Behr, Basel
Jean Opsomer, Ames
Andrea Pallini, Bologna
Andrea Pastore, Venezia
Cira Perna, Salerno
Wolfgang Polasek, Basel
Gunter Ritter, Passau
Roberto Rocci, Roma
Gilbert Saporta, Paris
Michael Schimek, Graz
Tonino Sclocco, Chieti
Roberta Siciliano, Napoli
Javier Trejos, San José
Gerhard Tutz, Munchen
Ralf Wagner, Bielefeld
Klaus-Dieter Wernecke, Berlin

Contents

Part I

Classification Methods with Applications

The STP Procedure as Overfitting Avoidance Tool in Classification Trees

Carmela Cappelli[1] and Francesco Mola[2]

[1] Dipartimento di Scienze Statistiche, Facoltà di Scienze Politiche Università degli Studi di Napoli, Via L. Rodinò, I-80138 Napoli, Italy
[2] Dipartimento di Economia Facoltà di Economia Università degli Studi di Cagliari, Viale S. Ignazio n.17, I-90123 Cagliari, Italy

Abstract. Among tree classification methods it will be shown how STP algorithm (Cappelli *et* al, 2002) is an useful tool of *overfitting avoidance*. The overfitting problem is the presence of "false" subdivision, that although reducing the total error, does not correspond to the true relationship between predictors and response variable. Classical methods based on overfitting evaluation by error terms seems not appropriated for our aim. It will be shown how STP procedure studing the dipendence between response variable and split variables, applied to both simulations and real examples can evaluate the presence of overfitting, preserving only significant subdivisions.

1 Introduction

Tree based classification and regression methodologies have proven to be a powerful nonparametric approach to data analysis leading to several methodological proposals and practical applications.
These methods can be summarized as follows: let Y be a response variable, which can take values either in the real line or in a set of previously defined classes, and let $X_1, X_2, \cdots, X_p$ be a set of covariates. The problem is to establish a relationship between Y and the covariates in the form of a binary tree.
A standard tree construction arises from a divide and conquer algorithm which recursively divides the data space into two subregions (subgroups of observations) according to a splitting criterion. At each step (tree node) the splitting criterion selects the best covariate and cut point associated with it. The goodness of split at a given node is measured in terms of achievable separability among the sibling nodes, i.e., data partitioning is led by the aim of making the distribution of the response variable as pure as possible within the nodes.

Because there is noise in the data, the structure resulting from the divide and conquer algorithm tends to be large and above all it overfits the data. As a consequence the tree has so many splits that its structure is difficult to interpret. Furthermore several splits, especially the terminal ones, are due to chance in the sense that they reflect particular features of the sample data rather than modeling real underlying relationships between the response and the covariates.
The problem of overfitting in tree based methods has been widely discussed in literature (see for example Quinlan, 1986). The standard approach to avoid overfitting (and besides complexity) is tree simplification which retrospectively prunes away some of the branches of the so called totally expanded tree.
Pruning methods proposed in literature (for a review see Mingers, 1989) are based on the evaluation of the classification/prediction accuracy.
By discarding branches on the basis of their error pruning improves the tree accuracy and its understandability but, overfitted branches are removed purely by chance.
In other words, since overfitting can be meant as "reporting false signals" then classical pruning cannot be seen as a tool to avoid it. In fact, it does not make use of any specific tool to verify whether signals are false.
Also, as a means to improve accuracy, pruning has been outperformed by new strategies (Breiman, 1996,1998; Freund and Shapire, 1998)based on resampling procedures. These have proven to produce more accurate classifiers/predictors, but come at the expense of loosing the structure of the tree, i.e., the knowledge of the splits.
When knowledge acquisition (Hand, 1998), is a concern, these strategies cannot be employed and since classical pruning also fails, a different approach to simplification is needed.
In the framework of the pioneer pruning method provided in the CART methodology (Breiman et al, 1984), a statistical test pruning procedure (STP) has been recently introduced by Cappelli *et* al (2002) both for calssification and regression. The test looks for that final tree structure which retains only those splits that

provide a significant contribution to explaining the relationships between the response and the covariates.
This paper focus on the case of calssification and it reports the results of examples and applications used to evaluate the performance of STP as an overfitting avoidance tool relative to the classical CART pruning method.
The paper is organised as follows: section 2 briefly describes the CART approach and the STP procedure for classification trees. Section 3 is devoted to some experiments.

2 Simplifying Trees

CART pruning method. CART approach is based on the concept of cost-complexity measure, which is a way to concentrate simultaneously on both the critical aspects of size and accuracy. For any node t and for the subtree T_t rooted at t the cost-complexity measure is defined as:

$$R_\alpha(t) = R(t) + \alpha, \tag{1}$$

$$R_\alpha(T_t) = R(T_t) + \alpha|\bar{T}_t|, \tag{2}$$

where $R(t)$ and $R(T_t)$ are the resubstitution error measures at node t and at subtree T_t respectively and $|\bar{T}_t|$ i.e. the number of leaves of the subtree.
The two measures equals if:

$$\alpha_t = \frac{R(t) - R(T_t)}{|\bar{T}_t| - 1}, \tag{3}$$

From the (3) we see that $\alpha(t)$ gives the gain in accuracy per terminal node in the given branch. Therefore, when the two measures coincide, there is no point in retaining the subtree which increases the size of the tree without improving its accuracy.
Although the complexity parameter takes a continuous range of values, the number of subtrees of the maximal tree is finite, thus the pruning process produces a finite sequence of subtrees. In other words, if T_α is the optimal tree that minimizes the cost-complexity measure for a given value of α this continues to be

optimal as α increases until a threshold $\alpha^{'}$ is reached and a new tree $T_{\alpha'}$ (with fewer terminal nodes) becomes the optimal one. This means that a tree is optimal for an interval of values of the complexity parameter and the sequence is generated by finding at each step the upper limit of the interval, i.e., the threshold given by the (3).

The algorithm consists of two stages: first the sequence of pruned subtrees is generated pruning the node that contributes the least gain in accuracy, then a single tree is selected. This is the one that minimizes the the error rate estimated by dropping test observations down the tree. This criterion is referred as $0-SE$ rule. Since the minimum is highly unstable, the choice of the tree with the minimum error rate might be arbitrary. An alternative criterion, therefore, is the $1-SE$ rule that selects the smallest tree whose error rate on the test set is within one standard error of the minimum.

The idea of generating a sequence of trees based on size and accuracy is undoubtfully appealing. The selection the optimum pruned subtree requires the space of all possible subtrees of the given totally expanded tree to be searched. This space is typically too vast; the cost complexity criterion represents a way to restrict the search to trees which are optimal.

On the other hand, the selection criterion which takes into account solely the aspect of accuracy, appears insufficient to the aim of detecting the "model" that underlies the data and explains the true relation between the response and the covariates.

In fact, the $0-SE$ rule tends to select large trees retaining spurious splits that accidentally reduce the error without being really related to the response, whereas, the $1-SE$ rule in most cases, results in the selection of a very small tree, showing a bias toward overpruning.

The STP procedure. Cappelli *et* al. (2001) Under the condition that the same measure, say I, employed both to grow and prune the tree, the complexity parameter can be proven to be:

$$\alpha(t) = \frac{1}{|\tilde{L}_t|} \sum_{l \in L_t} \Delta I(s^*, l), \tag{4}$$

i.e., as the mean of the reduction in the measure I induced by the best split s^* of each internal node l of the branch T_t, in the set L_t whose cardinality is $|\tilde{L}_t| = |\tilde{T}_t| - 1$.

This result is the starting point of the definition of a statistical testing procedure as the third stage in tree growing approach aimed to validate the pruning process, i.e., to find the reliable honest size tree.

By replacing the error rate with the Gini index of heterogeneity in the pruning process a different complexity parameter can be defined:

$$\beta(t) = \frac{1}{|\tilde{L}_t|} \sum_{l \in L_t} \Delta i(s^*, l) p(l), \tag{5}$$

where $p(l) = n(l)/n$ is the weight of node l; this complexity parameter is proven to be related to the χ^2 distribution as follows:

$$Jn\beta(t) \sim \chi^2_{(J-1)}. \tag{6}$$

The proposed procedure generates the sequence of pruned subtrees according to the CART method, but, using the Gini index instead of the error rate. Since for any internal node the complexity parameter represents the gain in accuracy induced by the corresponding subtree, the statistic (6) can be used to test the significance of this gain, i.e, to verify whether the corresponding branch should have been kept or pruned.

The testing considers an independent test set on whose observations the complexity parameters are re-computed. The values of complexity parameter increase each step on the training set, whereas on the test set they do not form an increasing sequence. Therefore the result is a single final tree, that, typically, does not correspond to any tree in the CART sequence.

3 Examples and Applications

The simplest way to show how the STP procedure copes with overfitting is to study how it performs in terms of size of the final tree and splitting variables included in when applied to data drawn from a known model.
To this aim we consider a very simple base model made up to provide insights into the problem of simplifying the tree. The model is as follows:

$$y_i = 0.01x_{i1} + 0.05x_{i2} + 5x_{i3} + \epsilon_i \qquad (7)$$

where x_{1i}, x_{2i}, x_{3i} are generated from the uniform distribution in the intervals [-1, 0], [0, 1] and [-0.5, 0.5] respectively and is drawn from a zero-mean homoschedastic normal:

$$\epsilon_i\ N(0, \sigma_\epsilon) \qquad i = 1, \ldots, n.$$

A dummy response variable y^* is created according to the following rule of thumb:

$$y^* = \begin{cases} 1 \text{ if } y \leq 0, \\ 2 \text{ if } y > 0 \end{cases}$$

Given the way the model is constructed, covariates x_1 and x_2 play no real role in determining the response which depends exclusively on covariate x_3.
If no error term is added to the model, a classification tree would be characterized by splits involving only this covariate. By adding the error term, the relation becomes less and less evident depending on σ_ϵ. and the
We consider $n = 1000$ observations, randomly generated according to model(7) where the error term standard deviation has been set $0.1, 0.3, 0.7, 2, 3$ respectively, resulting in 5 data sets.
Figure 1 shows the boxplots of covariate x_3 for the noiseless case ($\epsilon = 0$) and the extreme hypothesis on the error term standard deviation ($\sigma_\epsilon = 3$) against the two response classes.

The higher the error term standard deviation, the less strong the relation between the response and covariate x_3; going from

the noiseless data to the extreme hypothesis ($\sigma_\epsilon = 3$) we see that the boxes in the plots are almost overlapped. In the application the data sets are randomly split into a training set (70%) and a test set (30For each data set the classification tree is built using the Gini index as impurity measure to evaluate the goodness of split, i.e. to grow the tree. The resulting totally expanded tree is pruned according to both the classical CART the STP procedure. The former creates the sequence of increasingly pruned subtrees aiming at minimize the error rate. The latter uses the Gini index in creating the sequence and then it validates at
A tree is grown also for the noiseless data; only a single split along covariate x_3 fits the training set perfectly.
The results are summarized in table 2 that, for each case, reports the number of terminal nodes and the splitting variables of the totally expanded tree, the final CART tree and the STP tree, respectively. In the STP procedure a significance level 0.01 is considered.

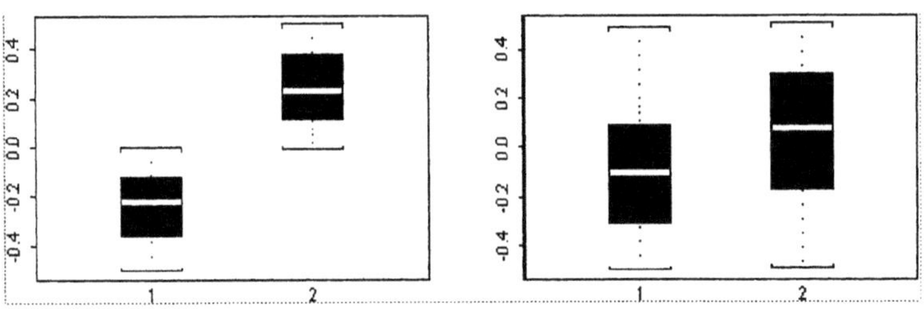

Fig. 1. The boxplots of covariate x_3 for $\epsilon = 0$ and $\sigma_\epsilon = 3$

As a general observation, trees are highly affected by noise. For $\sigma_\epsilon = 0.1$ the procedure grows a tree with 7 terminal nodes whereas in the noiseless case a single split on covariate x_3 is able to separate the two classes. As a matter of fact, one-depth trees, i.e., trees able to provide a satisfactory fit to the data with a single split, are not uncommon in practical applications, as shown by Holte (1993).
Increasing the error results in maximal trees more and more larger. This behavior is the consequence of the divide and conquer strategy. This, in the effort to model all the data, i.e, to make nodes

Table 1. Size and splitting variables included in the maximal trees, the CART trees ($0 - SE$ rule) and the STP trees.

	Maximal trees		CART trees		STP trees	
Error term	*Size*	splitting variables	*Size*	splitting variables	*Size*	splitting variables
$\epsilon_i \sim N(0, 0.1)$	7	x_1, x_2, x_3	4	x_1, x_3	2	x_3
$\epsilon_i \sim N(0, 0.3)$	25	x_1, x_2, x_3	13	x_1, x_2, x_3	3	x_3
$\epsilon_i \sim N(0, 0.7)$	64	x_1, x_2, x_3	19	x_1, x_2, x_3	3	x_3
$\epsilon_i \sim N(0, 2)$	95	x_1, x_2, x_3	16	x_1, x_2, x_3	2	x_3
$\epsilon_i \sim N(0, 3)$	125	x_1, x_2, x_3	23	x_1, x_2, x_3	3	x_3
$\epsilon_i \sim N(0, 0.01x_2 + 0.05\|x_1\|)$	12	x_1, x_2, x_3	8	x_1, x_2, x_3	2	x_3

purer and purer, ends involving splits on covariates having no explanatory power respect to the response, namely x_1 and x_2. CART pruning reduces strongly the size of the totally expanded tree but since it is not meant to address overfitting, it is affected by it resulting anyway in overly large trees which include splits on the two "weak" covariates.
On the contrary STP being based on the statistical evaluation of the dependence of the response from the splitting variable is able to detect meaningfulness splits and to avoid them. The size of the final trees is the same or slightly higher than that of the noiseless tree retaining only the higher levels splits on covariate x_3; even in the most extreme case, when the noise in the data creates almost a situation of inseparability between the two response classes, the test is able to detect the relation between the response and covariate x_3. Note that a further increase in the error term standard deviation, causes the tree methodology to fail and the first split is along covariate x_1.
An example with heteroscedastic error is also considered. Although the maximal tree shows a modest size (12 terminal nodes) it includes splits on covariates $x1_1$ and x_2 in the upper levels of the tree; the final tree selected by means of the $0 - SE$ rule with 8 terminal nodes is comparatively large. According to the STP procedure only the first split (covariate x_3) turned out to be significant.

Applications to real data sets are considered as well. The data are: the *Wisconsin diagnostic breast cancer* data (wdbc), from the UCI machine learning repository (http://www.ics.uci.edu/~mlearn/MLRepository.html) the *credit* data and the *Kyphosis* data from the library of the software SPAD and SPLUS respectively. The wdbc data consists of 569 cellular samples, on which a binary response variable (malignant and benignant) and 30 real-valued features of the cell nuclei, were recorded. The credit data consists of 468 clients of a bank; the response is binary, reliable clients (those who paid their debts) and unreliable ones, there are 11 nominal covariates. The kyphosis data has 81 instances representing children who have had corrective spinal surgery. The outcome is a binary variable (whether a postoperative deformity is "present" or "absent"), to be predicted based on three numeric variables.
Table 2 reports the results of the comparison. In this case, since the true underlying tree model is unknown, to evaluate the outcome of the CART selection rules and of the STP procedure, the AIC and the BIC model selection criteria are considered. Trees can be represented as contingency tables that cross-classifies the terminal nodes and the response classes. For these tables the $AIC =$ -2(maximised likelihood -n.of parameters) ranks the models in a way that is equivalent to use $G^2 - 2(df)$ where G^2 is the loglikelihhod ratio test statistic; the criterion $BIC = G^2 - log(n)(df)$ takes sample size into account.

Table 2. Size and values of of the AIC and BIC criteria for the CART trees and the STP trees.

	Wdbc			**Credit**			**Kyphosis**		
methods	*Size*	*AIC*	*BIC*	*Size*	*AIC*	*BIC*	*Size*	*AIC*	*BIC*
0-SE	12	444.804	401.463	33	240.904	119.155	10	28.900	10.513
1-SE	4	392.053	380.232	10	157.440	123.303	7	27.730	15.472
STP	4	388.738	376.917	8	125.913	103.155	2	10.450	8.407

The STP procedure outperforms the CART selection rules, achieving considerably smaller values of the criteria indicating better fits to the data. These results, therefore, confirm that the selection rules proposed in the CART are somewhat arbitrary from the point of view of the fits provided by the tree model to the data. Note that the STP selects for all the examples smaller trees compared to CART but for the wdbc data. This case is particular interesting since the $1 - SE$ tree and the STP tree have the same size but they differ for a single split. Nevertheless this slight difference in the structure of the tree causes a difference in the fit provided to the data showing that minimizing the error can be a misleading strategy.

4 Concluding Remarks

Statistical tests for pruning trees was proposed as a top down stopping rule to stop growing one depth branch (Mola e Siciliano, 1994; Lanubile e Malerba, 1998) or to prune retrospectively one depth branch (Mingers, 1989). The STP differs significantly from these since it takes advantage of the intuitive and appealing strategy proposed in the CART methodology combining it with a desirable statistical validation. It searches in the space of the CART-like pruned trees investigating the dependence relation of the response from the splitting variables. The experiments show that the STP is a promising approach to detect false relations reported in the growing phase, i.e., to overfitting avoidance. This is particularly needed when the knowledge of the explanatory splits, is a concern.

References

AGRESTI, A. (1990). *Categorical data Analysis.* Wiley Series in probability and Mathematical Statistics.

BREIMAN, L., *et* al (1984). *Classification and Regression Trees*, Wadsworth, Belmont CA.

BREIMAN, L. (1996): Bagging predictors. *Machine Learning, 24*, 123-140.

BREIMAN, L. (1998): Arcing classifiers. *The Annals of Statistics, 26 (3)* , 801-849.

CAPPELLI, C.,*et* al (2002): A Statistical Approach to Growing a Honest Reliable Tree. *Computational Statistics and Data Analysis, 38*, PAGINE.

FREUND, Y. , SHAPIRE, R. E. (1996): Experiments with a new boosting algorithm. In: *Machine Learning: Proceedings of the XIII International Conference*, 148-156.

LANUBILE, A., MALERBA, D. (1998) Induction of Regression Trees with Regtree. In: *Book of Short Papers of the Biannual Meeting of the Classification Group of SIS* , 253-256, Pescara.

MINGERS, J. (1989): An Empirical Comparison of Pruning Methods for Decision Tree Induction. *Machine Learning, 4*, 227-243.

MOLA, F., SICILIANO, R. (1994): Alternative strategies and CATANOVA testing in two-stage binary segmentation. In: Diday E. et al Eds.: *New Approaches in Classification and Data Analysis*, Heidelberg: Springer Verlag, 316-323.

HAND, D.J. (1998). *Construction and Assessment of Classification Rules*, London: Wiley.

HOLTE R.H. (1993): Very simple classification trees perform well on most commonly used data sets. *Machine Learning*, *11*, 63-91.

QUINLAN, J.R. (1985).The effect of noise on concept learning. In: R.S Michalski, J. Carbonell and T Mitchell (eds), *Machine Learning : An Artificial Intelligence Approach*, *2*, Los Altos : Morgan Kaufman, 149-166.

A Modal Symbolic Pattern Classifier

Francisco de A.T. de Carvalho[1], Renata M.C.R. de Souza[1], and Rosanna Verde[2]

[1] Centro de Informatica - CIn / UFPE, Av. Prof. Luiz Freire s/n - Cidade Universitaria CEP 50740-540 Recife-PE BRASIL

[2] Dipartimento di Strategie Aziendali e Metodologie Quantitative - Seconda Universita di Napoli, Piazza Umberto I, 81043 Capua, Italia

Abstract. This paper aims to presenting a new algorithm to classify symbolic data. The input data for the learning step is a set of symbolic objects described by symbolic interval (or set-valued) variables. At the end of the learning step, each group is represented by a (modal) symbolic object which is described by symbolic histogram (or bar-diagram) variables. The assignment of a new observation to a group is based on a dissimilarity function which measures the difference in content and in position between them. The difference in position is measured by a context free component whereas the difference in content is measured by a context dependent component. To show the usefulness of this modal symbolic pattern classifier, a particular kind of simulated images is classified according to this approach.

1 Introduction

In many domains, huge sets of data are recorded in large databases. A basic step of the *Symbolic Data Analysis* (Diday (2000)) consists in summarizing these data and extract new knowledge from them by means of *symbolic data*. Symbolic data are more complex than the standard one as they contain internal variation and they are structured data. The main goal of Symbolic Data Analysis is to apply new tools in this extracted knowledge in order to extend *Data Mining* in *Knowledge Mining* (Diday (2000)). Therefore, an extension of data analysis and statistical methods to the symbolic data is necessary.

Ichino *et al.* (1996) introduced a symbolic classifier as a region oriented approach based on Boolean symbolic data. In the learning step, each group is described by a disjunction of Boolean symbolic objects, which are obtained by using as tools of generalisation an approximation of the *Mutual Neighbourhood Graph* (MNG) and a Boolean symbolic operator (join). At the end of this step, the classifier furnishes a complete and discriminant description of the examples of each group. In the classification step, the allocation of an observation to a group is based on a matching function between an usual description and a Boolean symbolic description, which describes a group.

Souza *et al.* (1999) and De Carvalho *et al.* (2000) have proposed a new approach in order to get a MNG approximation and to reduce the complexity of the learning step, without loss of the classifier performance (measured by the error rate of classification). This new approach has made possible to study a special kind of simulated images (SAR images). At the classification step, new context free similarity and dissimilarity measures have been introduced in order to compare a standard description of an individual with a Boolean symbolic description of a group.

This paper aims to presenting an algorithm for building a symbolic classifier where each group is represented by a modal symbolic description. In such framework, a compromise dissimilarity function is proposed. It is based on two components: a location and content ones. The first component is supposed independent from the context while the second one is context dependent.

2 Symbolic Object Descriptions

In classical data analysis, an individual is described by a row of a data matrix (the columns are the variables), which assumes only one value from its domain. According to the kind of domain, a variable may be quantitative (discrete or continuous) or qualitative (ordinal or nominal). However, in the real world the recorded information are often too complex to be described by usual data. It is why different kinds of symbolic variables and symbolic data have been introduced.

A symbolic variable is *set-valued* if, for an object (individual or class), it takes a subset of categories from its domain. In particular, an interval variable takes, for an object, a continue set of values in a sub-set of $\Re$, whereas a modal variables take, for an object, a non-negative measure (a frequency or a probability distribution or a system of weights). If this measure is specified in terms of *histogram*, the modal variable is called *histogram variable*, whereas if the measure is specified by a *bar diagram* the modal variable is called *bar diagram variable* (Bock (2000)).

In the context of the Symbolic Data Analysis, a symbolic description of an object is a symbolic data vector whose columns are symbolic variables. These symbolic descriptions are collected in a Symbolic Data Table, which are more complex then the usual one because each cell contains more than a single value. Some weights can be associated to such intervals as well as logical rules and taxonomies can be also considered on the symbolic descriptors (Diday (2000)).

Example 1. A segment (set of pixels) may have as symbolic description the symbolic data vector $d_s = ((0.7[70,80[), (0.3[90,120[))$, where the characteristic grey level, which takes values in $[70,80[$ and $[90,120[$, is modelled as a *histogram* (or modal) variable, and 0.7 and 0.3 are the relative frequencies of the two intervals of values.

3 A Modal Symbolic Classifier

In this section, we are going to describe one of the versions of the symbolic classifier based on modal descriptions that we have implemented.

3.1 Learning Step

Two steps constitute the *learning process*: *pre-processing* and *generalization*.

Pre-processing. The aim of this step is to represent each segment as a modal symbolic description. The segments, obtained from a segmentation algorithm, are the input of the learning step. Each segment is constituted by a set of pixels, on which different levels of grey are measured. For each segment s we consider an interval I_s representing the range of the grey level of the pixels.

Example 2. Table 1 shows examples of some segments from two distinct regions of an image. In this table, a segment is described by an interval variable which allows to represent the range of the *grey levels* of the set of pixels belonging to it.

Table 1. Segments described by a symbolic interval variable

Segments (set of pixels)	Grey level (y_1)	Region
Seg_1	[10,30]	1
Seg_2	[25,35]	1
Seg_3	[90,130]	2
Seg_4	[125,140]	2

From the intervals describing the segments, we create a new set of intervals $I_T = \{I_j\}_{j \in J}$, such that $I_j \cap I_{j'} = \emptyset, \forall j, j' \in J$, with $j \neq j'$, as follows: at first, we take the set of values formed by every bound (lower and upper) of all the intervals associated to the initial segments. Then, such set of bounds is sorted in a growing way. Every interval of I_T is defined by two successive limits of this orderly set.

Example 3. From Table 1, the new set of intervals is $I_T = \{I_1, I_2, I_3, I_4, I_5, I_6, I_7\}$ where $I_1 = [10, 25[, I_2 = [25, 30[, I_3 = [30, 35[, I_4 = [35, 90[, I_5 = [90, 125[, I_6 = [125, 130[$ and $I_7 = [130, 140]$.

In this way, if I_s is the interval which describes the segment s, then to every $I_j \in I_T$ it is verified one of the following conditions: $I_j \cap I_s = \emptyset$ or $I_j \subseteq I_s$. Let $J_s = \{j \in J / I_j \subseteq I_s\}$. A segment s can be represented by means of a modal symbolic description, determined as follows:

$$d_s = ((w_{ks}, I_k)_{k \in J_s}), w_{ks} = \frac{|I_k|}{\sum_{k \in J_s} |I_k|} \quad (1)$$

where $|I_k|$ is the range of the interval I_k.

Example 4. Table 2 shows the segments of Table 1 described by a symbolic *histogram* variable.

Table 2. Segments described by a symbolic histogram variable

Segments (set of pixels)	Grey level (y_1)	Region
Seg_1	((0.75[10,25[),(0.25[25,30[))	1
Seg_2	((0.50[25,30[),(0.50[30,35[))	1
Seg_3	((0.88[90,125[),(0.12[125,130[))]	2
Seg_4	((0.33[125,130[),(0.67[130,140]))	2

Generalization. This step aims to represent each Region as a modal symbolic object. The symbolic description of each region is a generalization of the modal symbolic description of its segments.

Let $S = \{s_n\}_{n \in N}$ be a set of segments belonging to the same Region r, and $D = \{d_n\}_{n \in N}$ their modal symbolic descriptions. Let $d_n = \left((w_{jn} I_j)_{j \in J_{s_n}}\right)$ be the modal symbolic description of s_n. The Region r is similarly represented by a modal symbolic description as follows:

$$d_r = \left((w_{kr}, I_k)_{k \in \bigcup_{n \in N} J_{s_n}}\right), \text{ with } w_{kr} = \frac{1}{\#N} \sum_{n \in N} w_{kn} \tag{2}$$

where #N is the number of segments belonging to the group r. For every segment s_n there is an associated interval I_{s_n}. We also associate to the Region r an interval I_r which is defined as: $I_r = \bigcup_{m \in M} I_m$, where $M = \{j \in J / I_j \subseteq I_{s_n} \text{ for at least one } n \in N\}$.

Example 5. Table 3 shows the modal symbolic descriptions of the Regions:

Table 3. Regions described as a modal symbolic description

Region	Grey level (y_1)
1	((0.38[10,25[),(0.37[25,30[),(0.25[30,35[))
2	((0.44[90,125[),(0.23[125,130[),(0.33[130,140]))

3.2 Allocation step

The affectation of a new observation (segment) to a group (region) is based on a dissimilarity function, which compares the symbolic description of the new segment and the symbolic description of the region. The dissimilarity function, which will be described later on, measures the difference in contents and in position between the segment and the group. This dissimilarity is a hybrid function: the difference in position is measured by a context free component, whereas the difference in content is measured by a context dependent component.

Two steps also constitute the allocation process: *pre-processing* and *assignment*.

Pre-processing. The aim of this step is to represent the new segment by a modal symbolic description. In this paper, this is achieved by associating to its interval a weight equal to 1.

Assignment. Let x be a new segment with description $d_x = ((w_{kx}I_k)_{k \in K})$ and r be a Region with description $d_r = ((w_{jr}I_j)_{j \in J})$. As we know, the interval I_x, associated to x, is such that: $I_x = \cup_{k \in K} I_k$ and $I_k \cap I_{k'} = \emptyset, \forall k, k' \in K$ where $k \neq k'$. Let L_x be the set of lower bounds of the intervals $I_k, k \in K$, i.e., $L_x = \{L_k | I_k = [L_k, U_k], k \in K\}$, and let $L_r = \{L_j | I_j = [L_j, U_j], j \in J\}$. Let U_x be the set of upper bounds of the intervals $I_k, k \in K$, i.e., $U_x = \{U_k | I_k = [L_k, U_k], k \in K\}$, and let $U_r = \{U_j | I_j = [L_j, U_j], j \in J\}$.

The *context free component* of the dissimilarity function is defined by:

$$\phi_{cf}(x, r) = \frac{|\overline{I}_x \cap \overline{I}_r \cap (I_x \oplus I_r)|}{|I_x \oplus I_r|} \tag{3}$$

where $I_x \oplus I_r$ is the *join-interval* of the two intervals I_x and I_r (Ichino et al (1996)), which is defined as

$$I_x \oplus I_r = [min\{min(L_x), min(L_r)\}, max\{max(U_x), max(U_r)\}] \tag{4}$$

The functions *min* and *max* furnish, respectively, the lowest and the greatest value among a set of values.

The *context dependent component* of the dissimilarity function is defined by:

$$\phi_{cd}(x,r) = \frac{1}{2}\left(\sum_{k\in K} \frac{|I_k \cap \overline{I}_r|}{|I_k|} w_{kx} + \sum_{j\in J} \frac{|I_j \cap \overline{I}_x|}{|I_j|} w_{jr}\right) \quad (5)$$

The global dissimilarity function is given by the sum of the two components:

$$\phi(x,r) = \phi_{cf}(x,r) + \phi_{cd}(x,r) \quad (6)$$

4 The Monte Carlo Experiences

Synthetic Aperture Radar (SAR) is a system that possesses proper illumination and produces images with high capacity to discriminate objects. It uses coherent radiation, generating images with speckle noise. Data SAR possesses a random behaviour that has been usually explained by a multiplicative model (Frery et al. (1997)). This model considers that the observed return signal Z is a random variable defined as the product of two random variables: X (the terrain backscatter) and Y (the speckle noise).

A special kind of SAR simulated image will be classified according to the approach described on the previous section (Souza et al (1999), De Carvalho et al. (2000)). The process to obtain simulated images consists in creating classes of idealized images (a phantom, see Figure 1), and then, in associating to every classes a particular distribution.

Different kinds of detection (intensity or amplitude format) and types of regions can be modelled by different distributions associated to the return signal. Concerning the region types, the homogeneous (e.g. agricultural fields), heterogeneous (e.g. primary forest) and extremely heterogeneous (e.g. urban areas) are considered in this work. According to Frery et al. (1997) we assume that the return signal in amplitude case has the *square root of a Gamma* distribution, the K-Amplitude distribution and the G0-Amplitude distribution in homogeneous, heterogeneous and extremely heterogeneous areas, respectively.

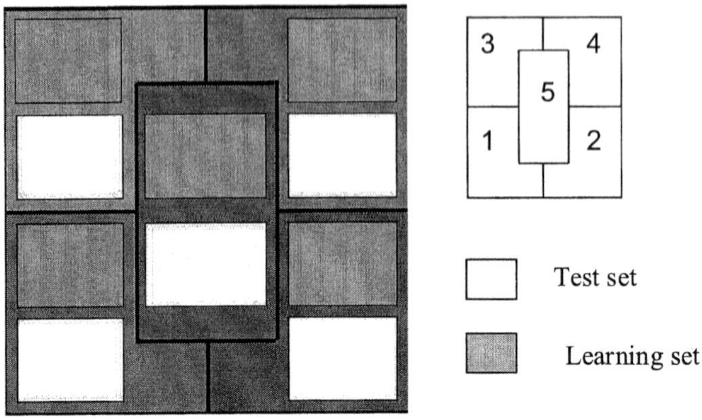

Fig. 1. Phantom with five regions

We generate the distributions associated to each class in each situation by using an algorithm for generating gamma variables (see Figure 2).

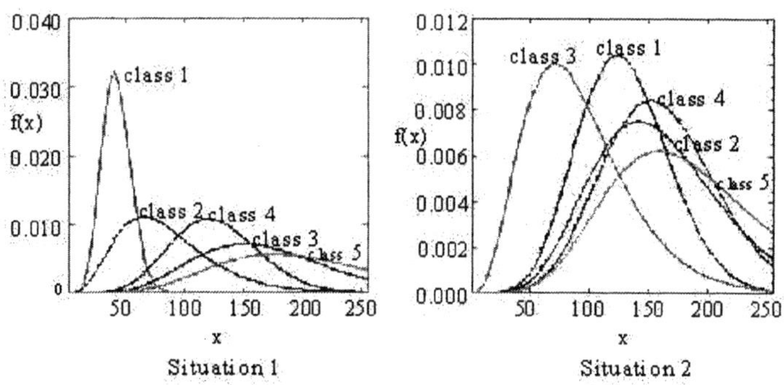

Fig. 2. Densities of SAR images

The parameters of these distributions are specified in the Table 4 and 5. Moreover, the *Lee filter* (Lee (1981)) was applied to the data before segmentation in order to decrease the effect of the speckle noise (see Figure 3).

The segmentation was obtained using the region growing technique (Jain (1989)), based on the T-student test (at the 5% significance level) for the merging of regions. The Monte Carlo experience was performed for images of size 512×512 and considering

Table 4. Distributions and parameters for situation 1

Regions	Distributions	μ	α	β	λ	γ
1	$\Gamma^{1/2}$	42	-	1916	-	-
2	K-A	84	2	-	0.00023	-
3	K-A	168	8	-	0.00025	-
4	$\Gamma^{1/2}$	126	-	17249	-	-
5	G^0-A	210	-5	-	-	203987

Table 5. Distributions and parameters for situation 2

Regions	Distributions	μ	α	β	λ	γ
1	$\Gamma^{1/2}$	130	-	18361	-	-
2	K-A	150	5	-	0.00019	-
3	K-A	110	4	-	0.00028	-
4	$\Gamma^{1/2}$	160	-	27814	-	-
5	G^0-A	190	-5	-	-	166982

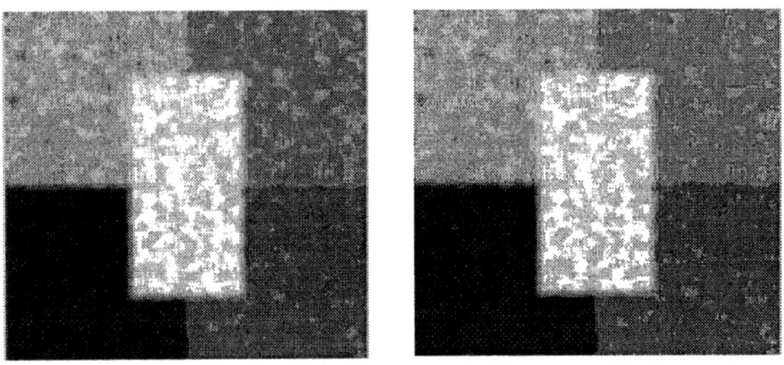

Situation 1 Situation 2

Fig. 3. Filtered images

two situations, ranging from moderate to great difficulty of classification. 100 replications were obtained with identical statistical properties.

The performance of the classifier was measured through the error rate of classification, which was estimated by the Monte Carlo method. The estimated error rate of classification corresponds to the average of the error rates found for these replications. The results were, in mean, 2.05% and 17.27% for each considered situation. In the Table 6, the confidence interval was constructed using the T-student test.

Table 6. Results of the Monte Carlo experiments

SAR Images	Average error rate (%)	Standard deviation	Confidence interval ($\alpha = 5\%$)	
			Lower Bound	Upper Bound
Situation 1	2.05	0.0306	1.98	2.11
Situation 2	17.27	0.1173	17.08	17.50

5 Concluding Remarks

We have presented a symbolic classifier based on modal symbolic descriptions. The usefulness of this classifier is corroborated by the good error rate under situations ranging from moderate to great difficulty of classification in the framework of a Monte Carlo experience of simulated images. This approach improved those who were presented by De Carvalho et al. (2000), Souza et al (1999) and Ichino et al. (1996) in computational time complexity. Other approaches for the construction of a symbolic classifier based on symbolic modal descriptions are in progress.

Acknowledgements. This paper is supported by grants from CNPq (Brazil) and MURST Confin 2000: *Modellizzazione di dati complessi attraverso tecniche di classificazione numerico-conceituale* (coord. R. Verde).

References

BOCK, H. H. (2000) Symbolic Data. In: Bock, H. H. and Diday, E. (Eds.): *Analysis of Symbolic Data*, Springer, Heidelberg, 39–53.

DE CARVALHO, F.A.T., ANSELMO, C.A.F., SOUZA, R.M.C.R. (2000) Symbolic approach to classify large data sets. In: Kiers, H.A.L. et al (Eds.): *Data Analysis, Classification, and Related Methods*, Springer, Heidelberg, 375–380.

DIDAY, E. (2000) Symbolic Data Analysis and the SODAS Project: Purpose, History, Perspective. In: Bock, H. H. and Diday, E. (Eds.): *Analysis of Symbolic Data*, Springer, Heidelberg, 1–23.

FRERY, A.C., MUELLER, H.J., YANASSE, C.C.F., SAN'ANNA, S.J.S. (1997) A model for extremely heterogeneous clutter. *IEEE Transactions on Geoscience and Remote Sensing*, *1*, 648–659.

ICHINO, M., YAGUCHI, H., DIDAY, E. (1996) A fuzzy symbolic pattern classifier. In: Diday, E. et al (Eds.): *Ordinal and Symbolic Data Analysis*, Springer, Heidelberg, 92–102.

JAIN, A. K. (1988) Fundamentals of Digital Image Processing. Englewood Cliffs. Prentice Hall International Editions.

LEE, J. S.(1981) Speckle analysis and smoothing of synthetic aperture radar images.*Computer Graphics and Image Processing, 17*, 24-32.

SOUZA, R.M.C., DE CARVALHO, F.A.T., FRERY, A.C. (1999) Symbolic Approach to SAR Image Classification. In: *Proceedings of the IEEE 1999 International Geosciences and Remote Sensing Symposium*, Hamburg, 1318–1320.

YAGUCHI, H., ICHINO, M., DIDAY, E. (1996) A knowledge acquisition system based on the Cartesian Space Model. In: Diday, E. et al (Eds.): *Ordinal and Symbolic Data Analysis*, Springer, Heidelberg, 113–122.

Proximity Measures Between Classification Trees

Rossella Miglio and Gabriele Soffritti

Dipartimento di Scienze Statistiche,
Via Belle Arti, 41 I-40126 Bologna, Italy
miglio@stat.unibo.it soffritt@stat.unibo.it

Abstract. Several proximity measures have been proposed to compare classifications derived from different clustering algorithms (Gordon (1999)). Few are the proposed solutions for the comparison of two classification trees. We have considered particularly two of these: one (Shannon and Banks (1999)) is a distance that measures the amount of rearrangement needed to change one of the trees so that they result in an identical structure, while the other (Miglio (1996)) is a similarity measure that compares the partitions associated to the trees taking into account their predictive power. In this paper we analyze features and limitations of these proximity measures and suggest a normalizing factor for the distance defined by Shannon and Banks; furthermore we propose a new dissimilarity measure that considers both the aspects explored separately by the previous ones.

1 Introduction

Classification trees represent non parametric classifiers that exploit the local relationship between the class variable and the predictors. They allow an automatic feature selection and a hierarchical representation of the measurement space. A typical segmentation procedure repeatedly splits the predictor space, generally, in two disjoint regions according to a local optimization criterion; in order to control the effective complexity of the model and to avoid overfitting a pruning procedure follows the growing step (Breiman et al. (1984)). A binary tree is the representation of such a sequence of binary partitions.

We can compare different classification trees in several situations. For example, samples of trees can be obtained from independent data sets containing the same predictors and class variables as in multicentric studies (Shannon and Banks (1999)). Furthermore, a multiplicity of these binary structures can derive from different greedy search algorithms used for identifying trees. Multiple versions of the same classifier can also be obtained over bootstrap datasets, and suitably combined in order to reduce the test set error (Breiman (1996), Freund and Schapire (1996)).

A way to compare a set of trees can be based on the use of measures that evaluate the similarity or distance between trees. A proximity measure can be used, in these situations, not only when we need to combine the observed set of trees into a single interpretable one (Shannon and Banks (1999), Cappelli and Shannon (2000)), but also to represent geometrically the group of trees via multi-dimensional scaling and to cluster this set of trees.

In this paper we analyze features and limitations of two proximity measures between classification trees: one is a distance that measures the amount of rearrangement needed to change one of the trees so that it has structure identical to the other (Shannon and Banks (1999)), while the other is a similarity measure that compares the partitions associated to the trees taking into account their relative predictive power (Miglio (1996)). We suggest a normalizing factor for the distance defined by Shannon and Banks; furthermore, we propose a new dissimilarity measure that considers both the aspects explored separately in the previous ones. The considered measures are then compared by analyzing some data sets in order to assess the capability of the new measure to detect differences in the structure and in the predictive power at the same time.

2 Comparing Different Tree Structures

The agreement of each tree T_i, $i = 1, \ldots, v$, with another tree T_j, $j \neq i$, can be assessed on the subset D_j used to grow T_j. On each data set D_j, $j = 1, \ldots, v$, the corresponding tree T_j offers a description of the data, which is as close as possible to this subset. The agreement of T_i with T_j is then a measure of how well the description of D_i applies to D_j. As a result, the stability measured in this way expresses the degree to which a description of a data set D_i can be generalized to the others. This proposal comes from

Ciampi and Thiffault (1988), who select the tree which best agrees with the others.

Two trees can be compared on a dataset D by measuring similarity between the two classification rules that the trees represent. This problem has its own counterpart in the comparison of two partitions of the same set of objects in the context of cluster analysis.

First we consider a similarity measure between classification trees.

Let T_i and T_j be two different trees, with H and K leaves respectively, that can be used to classify a set of n observations. We may label from one to H the leaves of T_i and from one to K the leaves of T_j and form the matrix

$$\mathbf{M} = [m_{hk}]\ h = 1, \ldots, H,\ k = 1, \ldots, K,$$

where m_{hk} is the number of objects which belong both to the h-th leaf of T_i and to the k-th leaf of T_j.

In analogy with a measure defined in Fowlkes and Mallows (1983) for the comparison of two hierarchical clusterings, a numerical measure of the degree of similarity between the two trees is then defined to be

$$B_{ij} = \frac{\sum_{h=1}^{H} \sum_{k=1}^{K} m_{hk}^2 - n}{\sqrt{P_i Q_j}}, \tag{1}$$

where

$$m_{h0} = \sum_{k=1}^{K} m_{hk};\ m_{0k} = \sum_{h=1}^{H} m_{hk};\ P_i = \sum_{h=1}^{H} m_{h0}^2 - n;\ Q_j = \sum_{k=1}^{K} m_{0k}^2 - n.$$

The B_{ij} value is bounded between 0 and 1. B_{ij} is equal to 1 when $\mathbf{M}$ has exactly $\min(H, K)$ elements that differ from 0, that means the leaves in each tree correspond completely. On the contrary, B_{ij} equals zero when each m_{hk} is 0 or 1, so that every pair of objects that appear in the same leaf in T_i are assigned to different leaves in T_j. B_{ij} is also related to the sum of those pairs that have matching leaves assignment.

If we want to compare two trees from the point of view of their associated partitions and predictions, it seems natural to modify this measure of similarity. The tree, in fact, represents a classification rule and the comparison of two partitions derived by different trees should take into account whether the elements of the partitions assign the same class label to each object. Let $c_{hk} = 1$ if the h-th leaf of T_i has the same class label as the k-th leaf of T_j, and $c_{hk} = 0$ otherwise. The similarity measure (1) has been modified as follows (Miglio (1996)):

$$B(T_i, T_j) = \frac{\sum_{h=1}^{H} \sum_{k=1}^{K} m_{hk}^2 c_{hk} - \sum_{h=1}^{H} \sum_{k=1}^{K} m_{hk} c_{hk}}{\sqrt{P_i Q_j}}, \tag{2}$$

where the numerator counts the pairs of objects which are classified together in both partitions and that are also assigned to the same class.

Shannon and Banks (1999) proposed a novel method for combining a sample of classification trees using maximum likelihood estimation which results in a single, interpretable tree.

Their method requires the computation of distances between trees; thus they proposed also a distance metric between two trees that measures the amount of rearrangement needed to change one of the trees so that it has structure identical to the other. Specifically, the distance between trees T_i and T_j is defined by Shannon and Banks as

$$d(T_i, T_j) = \sum_r \alpha_r |S_{ijr}|, \tag{3}$$

where S_{ijr} denotes the set composed by discrepant r-paths (paths of length r found in only one of the two trees), and $\alpha_r = \alpha(r)$ is a weight function that depends only on r.
The weight function can be selected to penalize structural differences between trees based on where these occur. For example, a constant weight function ($\alpha_r = 1$) does not distinguish between differences occurring early or late in the trees, while the function $\alpha_r = 1/r$ penalizes early discrepancies more heavily than those occurring later in the tree. A different choice of this component

allows to modify the distance metric to take into account expert opinions.

The value of $d(T_i, T_j)$ cannot be easily interpreted on its own to describe the intensity of the distance (such as to compare the distance between different pairs of trees) since it is highly dependent on the number of paths present in T_i and T_j and on the lengths of each of them. To overcome this limitation, the value of $d(T_i, T_j)$ should be divided by a normalizing factor max$d(T_i, T_j)$ obtained computing the distance between a pair of trees having exactly the same number of paths of (T_i, T_j) with the same lengths, but constructed so as to result discrepant. This can be easily obtained by choosing a splitting variable for the root node in T_i different from the one in T_j. In this way we obtain the following normalized measure:

$$D(T_i, T_j) = \frac{d(T_i, T_j)}{\max d(T_i, T_j)}. \tag{4}$$

Let us consider the characteristics of the previous measures. On one hand the similarity measure (2) compares the partitions associated to the trees taking into account their relative predictive power without taking into explicit consideration the variables identified as more discriminant. On the other hand, the distance (3) evaluates the difference between the structures of T_i and T_j with no regard for their associated partitions and predictions. When two classification trees have to be compared, both aspects (the structure and the predictive power) should be simultaneously considered. In fact, trees having the same distance with respect to their structures can show a very different predictive power. On the other hand, trees with the same predictive power can have very different structures. For this reason, we propose a new dissimilarity measure, which considers the structure and the predictive power at the same time. Its definition is the following:

$$\delta\left(T_i, T_j\right) = \sum_{h=1}^{H} \alpha_{ih}(1 - s_{ih})\frac{m_{h0}}{n} + \sum_{k=1}^{K} \alpha_{jk}(1 - s_{jk})\frac{m_{0k}}{n}, \tag{5}$$

where s_{ih} and s_{jk} are similarity coefficients whose values synthesize the similarities s_{hk} between the H leaves of T_i and the K leaves of T_j, computed as follows:

$$s_{hk} = \frac{m_{hk}c_{hk}}{\sqrt{m_{h0}m_{0k}}},\ h = 1, \ldots, H,\ k = 1, \ldots, K; \tag{6}$$

s_{hk} is equal to 0 when $m_{hk} = 0$ and/or $c_{hk} = 0$, and equal to 1 when $c_{hk} = 1$ and both the h-th leaf of T_i and the k-th leaf of T_j have exactly the same objects, that therefore are assigned to the same class. When T_i and T_j are equal, each leaf of T_i will result maximally similar to a particular leaf of T_j, and minimally similar to any other. For this reason, a suitable way of synthesizing the similarities s_{hk} seems to be choosing the maximum:

$$s_{ih} = \max\{s_{hk},\ k = 1, \ldots, K\}, \tag{7}$$

$$s_{jk} = \max\{s_{hk},\ h = 1, \ldots, H\}. \tag{8}$$

The coefficients α_{ih} and α_{jk} are dissimilarity measures: the first is computed between the h-th leaf of T_i and that particular leaf of T_j identified by the criterion (7), while the second between the k-th leaf of T_j and that particular leaf of T_i identified by the criterion (8). If the paths associated to these pairs of leaves are identical, α_{ih} (α_{jk}) is set equal to 0. On the contrary, α_{ih} (α_{jk}) has to be set greater than 0 on the basis of the values of q and p, which denote the length of the longest path and the level where the two paths differ from each other, respectively. Specifically, α_{ih} (α_{jk}) can be set equal to $q - p + 1$ when it is useless to distinguish between differences occurring early or late in the paths, or equal to $\sum_{l=p}^{q} \frac{1}{l}$ when early discrepancies have to be penalized. Furthermore, the introduction of the relative frequency of each leaf weights the discrepancies proportionally to the number of their observations. The maximum value of $\delta(T_i, T_j)$ can be reached when the difference between the structures of T_i and T_j is maximum and the similarity between their predictive powers is zero. The normalizing factor for $\delta(T_i, T_j)$ is thus equal to

$$\max \delta (T_i, T_j) = \sum_{h=1}^{H} \alpha_{ih} \frac{m_{h0}}{n} + \sum_{k=1}^{K} \alpha_{jk} \frac{m_{0k}}{n}, \tag{9}$$

where α_{ih} is set equal to q_{ih} (the length of the path from the root node to the h-th leaf of T_i) when a set of constant weights is considered, and equal to $\sum_{l=1}^{q_{ih}} \frac{1}{l}$ otherwise; α_{jk} is defined in an analogous way. The normalized version of the proposed dissimilarity is thus:

$$\Delta (T_i, T_j) = \frac{\delta (T_i, T_j)}{\max \delta (T_i, T_j)}. \tag{10}$$

3 Experimental Results

In this section we analyze the performance of the proposed measure by studying two different classification problems.

3.1 Threenorm example

The first example is a simulation study based on a two-class data set with 6-dimensional measurement vectors. Class 1 is drawn with equal probability from a unit multivariate normal with mean (a, ..., a) and from a unit multivariate normal with mean ($-a$, ..., $-a$). Class 2 is drawn from a unit multivariate normal with mean (a, $-a$, ..., a, $-a$), where $a = 2/(20)^{1/2}$. Sixty training samples of 500 observations each were generated and 10000 observations have been considered as test set. These trees were fit using a procedure like CART (Breiman et al. (1984)) implemented in GAUSS, with the Gini criterion for splitting and using the test set for pruning. Comparing pairs of the 60 trees obtained in this way, we have computed 1770 values for the measures (2), (4) and (10), considering both ways of weighting discrepancies for (4) and (10).

3.2 Breast cancer example

The second example is a real classification problem, where the objective is to correctly identify benign from malignant breast tumors (Mangasarian and Wolberg (1990)). It is a two-class problem. Each of the 699 instances consists of 9 cell attributes each of which is measured on a 1-10 scale plus the class variable. The Wisconsin Breast Cancer Database is in the UCI repository of machine learning databases (ftp.ics.uci.edu/pub/machine-learning-databases) and was obtained from the University of Wisconsin Hospitals. We generated 50 training sets of 600 observations, bootstrapping the original one, while the whole data set has been considered as a test set. Like in the previous example, also these trees were fit using a procedure like CART (Breiman et al. (1984)) implemented in GAUSS, with the Gini criterion for splitting and using the test set for pruning. Comparing pairs of the 50 trees on the test set we have obtained 1225 values for the measures (2), (4) and (10), considering both ways of weighting discrepancies for (4) and (10).

3.3 Results

Table 1 reports two Spearman correlation matrices between the considered measures, with respect to constant and decreasing weights, for the two examples. The comparison between measures has been done using an ordinal correlation measure instead of Pearsons coefficient because the empirical distributions of the proximities come closer to the bounds for some measures than others; for this reason, the relative ranks of the proximities should be compared instead of their absolute values. The values of B and D measures result to be not weakly correlated; in our opinion those correlations would be higher if we did not take into account every pair of trees which split the predictor space in very similar regions but using a different rule order. At least in the examples considered, the predictive power and the structure seem to be two aspects of a tree which are conceptually very different but which in practice result related to each other. For this reason we think that, when the dissimilarity between two trees is measured, the differences in predictive power and structure should be better evaluated by a single measure which is not influenced by pairs of trees discrepant only in the rule orders. In both cases the correlations between the proposed measure and the other two are greater

Table 1. Spearman correlations between the considered measures.

		Constant weights			Decreasing weights		
		B	D	Δ	B	D	Δ
Threenorm	B	1.000	-0.391	-0.792	1.000	-0.393	-0.808
	D		1.000	0.634		1.000	0.642
	Δ			1.000			1.000
Breast cancer	B	1.000	-0.557	-0.732	1.000	-0.553	-0.707
	D		1.000	0.805		1.000	0.864
	Δ			1.000			1.000

than the correlation between B and D; this is an expected result that we think is due to the capability of the new measure to detect differences in the structure and in the predictive power at the same time.

We have also studied how the distribution of the proposed measure changes when the similarity (2) grows up taking the dissimilarity (4) constant and, viceversa, when the dissimilarity (4) grows up taking the similarity (2) constant. Figure 1 reports the boxplots of the proposed measure distributions between the groups corresponding to the quintiles of the measure (2) within each quintile of (4) for the breast cancer example: with respect to the measure (4) it is possible to determine only three groups because third and fourth quintiles are equal. The obtained results show that fixing the dissimilarity (4) the mean of the proposed measure decreases according to the increasing of the similarity with respect to the predictive power; a similar but opposed result follows considering situations of constant similarity (2): the mean values of the proposed measure increase when the dissimilarity (4) decreases.

4 Concluding Remarks

In this paper we explore features and limitations of two proximity measures between classification trees. We also propose a new dissimilarity measure which overcomes some of the limitations highlighted.

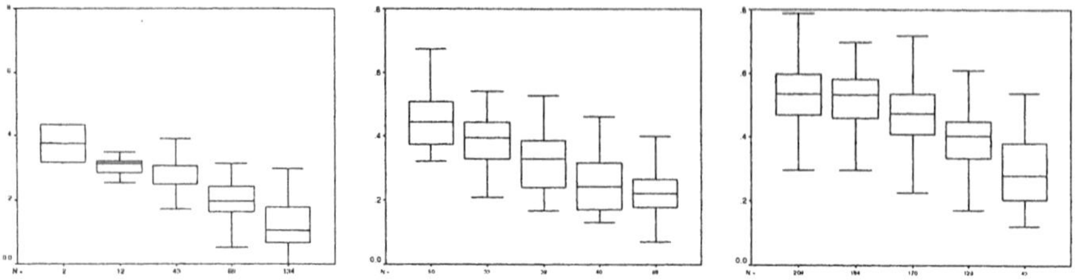

Fig. 1. Boxplots of the proposed measure distributions (constant weights).

Similarly to some measures of partition correspondence, each proximity measure between classification trees should also be corrected for chance (Hubert and Arabie (1985)) so as to take on some constant value under an appropriate null model.
We are presently examining the extension of the proposed measure to compare other types of trees (e.g.: regression trees, survival trees). We are also studying algorithms, based on the proposed measure, suitable for identifying a consensus tree which summarizes the information contained in different trees. When the trees are constructed on cross-validation samples, this strategy could represent an alternative to pruning techniques.

References

BREIMAN, L. (1996): Bagging Predictors. *Machine Learning, 24*, 123–140.

BREIMAN, L., FRIEDMAN, J.H., OLSHEN, R.A. and STONE, C.J. (1984): *Classification and Regression Trees.* Wadsworth, Belmont, California.

CAPPELLI, C. and SHANNON, W.D. (2000): An MLE Strategy for Combining Optimally Pruned Decision Trees. *COMPSTAT 2000*, 235 – 246, Phisica Verlag.

CAPPELLI, C., MOLA, F. and SICILIANO R. (2002): A Statistical Approach to Growing a Reliable Honest Tree. *Computational Statistics and Data Analysis, 38*, 285–299.

CIAMPI, A. and THIFFAULT, J. (1988): Recursive Partition in Biostatistics: Stability of Trees and Choice of the Most Stable Classification, *COMPSTAT 1988*, Physica Verlag.

FOWLKES, E.B. and MALLOWS, C. L. (1983): A Method for Comparing two Hierarchical Clustering. *Journal of the American Statistical Association, 78*, 553–584.

FREUND, Y. and SCHAPIRE, R. (1996): Experiments with a New Boosting Algorithm. *Machine Learning: Proceedings of the Thirteenth International Conference*, July 1996.

FREUND, Y. and SCHAPIRE, R. (1999): A Short Introduction to Boosting. *Journal of Japanese Society for Artificial Intelligence, 14*, 771–780.

GORDON, A. D. (1999): *Classification*, 2nd edition. Chapman and Hall/CRC, London.

HUBERT, L. and ARABIE, P. (1985): Comparing Partitions. *Journal of Classification, 2*, 193–218.

MANGASARIAN, O.L. and WOLBERG, W.H. (1990): Cancer Diagnosis via Linear Programming. *SIAM News, 23*, 1–18.

MIGLIO, R. (1996): *Metodi di Partizione Ricorsiva nell'Analisi Discriminante.* Dipartimento di Scienze Statistiche, Bologna.

SHANNON, W. D. and BANKS, D. (1999): Combining Classification Trees Using MLE. *Statistics in Medicine, 18*, 727–740.

Ordinal Classification Trees Based on Impurity Measures

Raffaella Piccarreta

Istituto di Metodi Quantitativi, Università L. Bocconi, v.le Isonzo 25, Milano.
E-mail: raffaella.piccarreta@uni-bocconi.it

Abstract. We introduce a new criterion for generating classification trees in the case when the response variable is ordered categorical. The framework is the well known CART methodology (Breiman et al. (1984)), which is actually the benchmark for the case when the response is nominal or numerical. Our criterion is obtained by measuring *impurity* within a node referring to a general measure of mutual dispersion (Gini (1954)), which can be applied to every variable, whatever its nature. We show that the two measures of impurity at the basis of CART in the case when the response is nominal or quantitative can be obtained as particular cases of the above mentioned measure. We also illustrate a property of our criterion, permitting the application of an algorithm recently proposed by Mola and Siciliano (1997, 1998) to fasten the process of growing the maximal tree.

1 Introduction

Consider the problem of predicting a response variable, Y, on the basis of a set of explanatory variables. The Classification And Regression Trees (CART) methodology of Breiman et al. (1984) addresses this problem by recursively partitioning the initial set of N cases into subsets being more and more homogeneous (with respect to the response variable). This is done by referring to the concept of *impurity* (heterogeneity) within a node and of *decrease of impurity* deriving by splitting a set of cases into two subsets. In particular, this kind of strategy has been defined in CART for the case when the response variable is nominal or quantitative. Our aim is to provide a new impurity measure within the CART splitting criterion in order to deal with the case when the levels of Y can be ordered. This is accomplished by referring to a measure of the dispersion of a variable introduced by Gini (1954). In particular, the impurity measures at the basis of CART (in the nominal and in the quantitative case) can be showed to be a particular case of the mentioned measure.

2 Classification Trees: CART Methodology

Consider a set of N cases, and suppose measurements are available on Q variables $X_1, X_2, \ldots, X_Q$. Suppose the N cases belong to G groups induced by the levels of a *response* variable, Y, which has to be predicted on the basis of the Q explanatory variables. Tree-structured classification addresses this problem by constructing a prediction rule in the form of a tree obtained by recursively partitioning the group constituted by all of the N cases into subsets, called *nodes*, through a sequence of binary splits. The classification tree approach of CART (Breiman et al. (1984)) consists into two main phases. In the *growing phase*, nodes are recursively split, and a sequence of nested trees is obtained, $\{T_i\} = \{T_1, T_2, \ldots, T_M\}$, having an increasing number of terminal nodes (i.e., nodes which are no longer split). This phase ends when the *maximal tree*, T_M, is obtained, a tree whose terminal nodes are either maximally Y-homogeneous (all cases are characterised by the same level of Y, and further splitting is not worthwhile) or maximally X-homogeneous (all cases are characterised by the same levels of the explanatory variables, and further splits are not possible) or are constituted by very few cases. In the *pruning phase* terminal nodes are sequentially merged to avoid over-fitting of the classification tree to the N cases used to grow it.

The focus of this work is on the *growing phase*. At the basis of this phase are the concepts of *impurity* (or heterogeneity) within a node and of *decrease of impurity* deriving by splitting a node into two sub-nodes. A node is split into two sub-nodes according to the values of one out of the explanatory variables. More precisely, all the possible sub-divisions (splits) which can be defined on the basis of the explanatory variables are evaluated, and the best split, s^*, is selected maximising the decrease in impurity. In CART two measures of impurity are defined to deal with the case when the response is categorical (nominal) or numerical. In the following section we will illustrate that these two measures can be seen as particular cases of a general measure of *mutual dispersion* of a variable. Our aim is to provide a criterion for generating the sequence $\{T_i\}$ in the case when Y is a categorical ordered variable.

This will be done by suitably "adapting" to the ordinal case the above mentioned measure of mutual dispersion.

3 Growing a Tree in the Ordinal Case

To motivate our proposal, we first of all briefly discuss about the concept of heterogeneity, playing a crucial role in CART methodology.

It is well known that heterogeneity can be measured (at least) in two way. Given a variable Y assuming G levels, $y_1, \ldots, y_G$ heterogeneity within a given node, t, can be measured by synthesising the (weighted) dissimilarity between each level of Y and a "pole" synthesising the distribution of Y:

$$S(t) = \sum_{g=1}^{G} d(y_g, \tilde{y}_t)\pi_t(g) \tag{1}$$

where $\pi_t(g)$ is the proportion of cases in node t characterised by the g-th level of Y. Of course, the choice of the measure of dissimilarity $d(.,.)$ and of the pole $\tilde{y}_t$ will depend upon the nature of the response variable.

Heterogeneity can also be evaluated by referring to the concept of *mutual dispersion* (Gini (1954)), a measure taking into account the dissimilarity of each level of Y from all the others and not from a synthesis (even if adequate) of Y:

$$H(t) = \sum_{g=1}^{G}\sum_{j=1}^{G} d(y_g, y_j)\pi_t(g)\pi_t(j) \tag{2}$$

Let us now focus attention on the cases when the response variable is nominal or quantitative. In the case when Y is nominal, the dissimilarity between two levels of Y can be measured by $d_N(y_g, y_j) = 1$ if $y_g = y_j$, 0 otherwise, and a proper synthesis of the distribution of Y is the *mode*, say y_{g^*} . In such situation (1) and (2) become:

$$S_N(t) = 1 - \pi_t(g^*) \quad \text{and} \quad H_N(t) = \sum_{g=1}^{G} \pi_t(g)[1 - \pi_t(g)] \tag{3}$$

$H_N(t)$ is the so called Gini concentration, the measure of heterogeneity CART procedure is based upon (in the nominal case).

As concerns the case when Y is a quantitative variable, one can choose $d_Q(y_g, y_j) = (y_g - y_j)^2$ and a possible synthesis of the distribution of Y is the mean, $\bar{y}_t$. It then follows:

$$H_Q(t) = 2\sum_{g=1}^{G}(y_g - \bar{y}_t)^2\pi_t(g) \propto S_Q(t) \tag{4}$$

The measure of impurity CART is based upon in the nominal and in the quantitative case is (2) – properly extended according to the nature of the response variable. Consider a node t, and suppose it is partitioned into two subgroups t_L and t_R by split s. In CART the *decrease in impurity* achieved when passing from t to t_L and t_R is defined as:

$$\Delta H(t, s) = H(t) - p_L H(t_L) - p_R H(t_R) \tag{5}$$

where p_L and p_R denote the proportion of cases in t_L and in t_R. In the nominal case, (5) reduces to the well known *Gini-Simpson criterion*:

$$\Delta H_N(t, s) = p_L p_R \sum_{g=1}^{G}[\pi_t(g|L) - \pi_t(g|R)]^2 \tag{6}$$

$\pi_t(g|L)$ and $\pi_t(g|R)$ being the proportion of cases characterised by the g-th level of Y in t_L and in t_R. In the quantitative case, one obtains:

$$\begin{aligned}\Delta H_Q(t, s) \propto \Delta S_Q(t, s) &= (\bar{y}_{t_L} - \bar{y}_t)^2 p_L + (\bar{y}_{t_R} - \bar{y}_t)^2 p_R \\ &= p_L p_R (\bar{y}_{t_L} - \bar{y}_{t_R})^2\end{aligned} \tag{7}$$

Notice that the decrease in impurity in (4) and (6) is given by the difference between *total heterogeneity* and *within nodes heterogeneity*, coinciding with the *between nodes heterogeneity*.

Consider now the problem of growing a classification tree when the G levels of Y are ordered. In this case, it is natural to require the terminal nodes to be constituted by cases characterised by levels of the response variable as much closest as possible. At this aim, it is necessary to refer to a measure of impurity within a node taking into account the ordinality of Y. Suppose a score, $\tilde{y}_g$, can be attached to each level of Y. Let then $d_{OS}(g,j) = |\tilde{y}_g - \tilde{y}_j|$ and let $\tilde{y}_t$ denote the *median score* within node t. It can be shown it is:

$$S_{OS}(t) = \sum_{g=1}^{G} |\tilde{y}_g - \tilde{y}_t| \pi_t(g), \tag{8}$$

$$H_{OS}(t) = \sum_{g=1}^{G} |\tilde{y}_{g+1} - \tilde{y}_g| F_t(g)[1 - F_t(g)] \tag{9}$$

$F_t(g)$ denoting the proportion of cases in t characterised by a level of Y lower than or equal to the g-th. Both measures are invariant in location, so that it suffices to define a measure of dissimilarity between two adjacent levels. Moreover, if one does not want to attach scores to the Y-levels, it is possible to set $\tilde{y}_g = g$. In this case, (8) becomes:

$$H_O(t) = \sum_{g=1}^{G} F_t(g)[1 - F_t(g)] \tag{10}$$

The criterion solely referred to in the ordinal case (see Fabbris (1997)) to evaluate a split s splitting node t into the sub-nodes t_L and t_R is based upon the measure of impurity $S_{OS}(t)$, but the decrease in impurity is evaluated as:

$$\begin{aligned} \Delta S_{OS}(t,s) &= |\tilde{y}_{t_L} - \tilde{y}_t| p_L + |\tilde{y}_{t_R} - \tilde{y}_t| p_R \\ &\neq S_{OS}(t) - p_L S_{OS}(t_L) - p_R S_{OS}(t_R) \end{aligned} \tag{11}$$

In a sense, notice that (10) is obtained by "adapting" criterion (6) to the ordinal case, but, differently from (6), can not be viewed as an extension of (5).

It is not the aim of this work to critic criterion $\Delta S_{OS}(t)$; our main point here is that impurity is measured by combining dissimilarities between each score and a (proper) synthesis of the distribution, the median. ¿From this point of view, this measure of impurity is not a measure of mutual dispersion such as the ones at the basis of the CART criterion (in the nominal and numerical case). Moreover, the decrease in impurity achieved by splitting a node into two sub-nodes is not obtained as the difference between the total heterogeneity within the node and the heterogeneity within the two obtained sub-nodes.

This is the reason why we suggest to construct classification trees measuring heterogeneity by referring to (8), which is a proper extension of the mutual dispersion measure to the ordinal case, and defining a criterion to select best splits on the basis of the difference between total and within heterogeneity.

The proposed criterion will constitute a possible alternative to the one solely referred to, enriching the class of instruments available for generating classification trees in the ordinal case.

Referring for simplicity to the case when $\tilde{y}_g = g$ (but results similar to those following hold also if one refers to the case when information about the score is available) we have:

$$H_O(t) = \sum_{g=1}^{G} F_t(g)[1 - F_t(g)] \tag{12}$$

$$\begin{aligned} \Delta H_O(t,s) &= H_O(t) - p_L H_O(t_L) - p_R H_O(t_R) \\ &= p_L p_R \sum_{g=1}^{G} [F_t(g|L) - F_t(g|R)]^2 \end{aligned} \tag{13}$$

$F_t(g|L)$ and $F_t(g|R)$ being the proportion of cases characterised by a level of Y lower than or equal to the g-th in t_L and t_R. The decrease in impurity in (12) is very similar to the criteria in (4) and (6): the so called *anti-end-cut* factor ($p_L p_R$) is combined with a measure of the dissimilarity between the two distribution of Y in t_L and t_R, such a measure depending upon the nature of the response variable (notice that in the quantitative case the difference between the distributions is measured by the distance between the means of Y within the two sub-nodes).

To appreciate differences between the trees generated by referring to criterion ΔS_O and ΔH_O respectively we now consider an application to a real data set.

4 An Application to Real Data

We consider a data set containing observations on 215 European banks. The response variable is the *world rank* of the bank (in 1995, source *Bankscope*); more precisely the response variable, Y assumes 8 levels, from greatest to lowest in degree ($Y = y_g$ if the rank is between $1000(g-1)$ and $1000g$). For each bank, information have been collected about its structure and performance (country, number of employees, total assets, funding equity, net income etc.). In Figure 1 and 2 the trees obtained with criteria ΔH_O and ΔS_O are reported (a darker colour corresponds to higher degree).

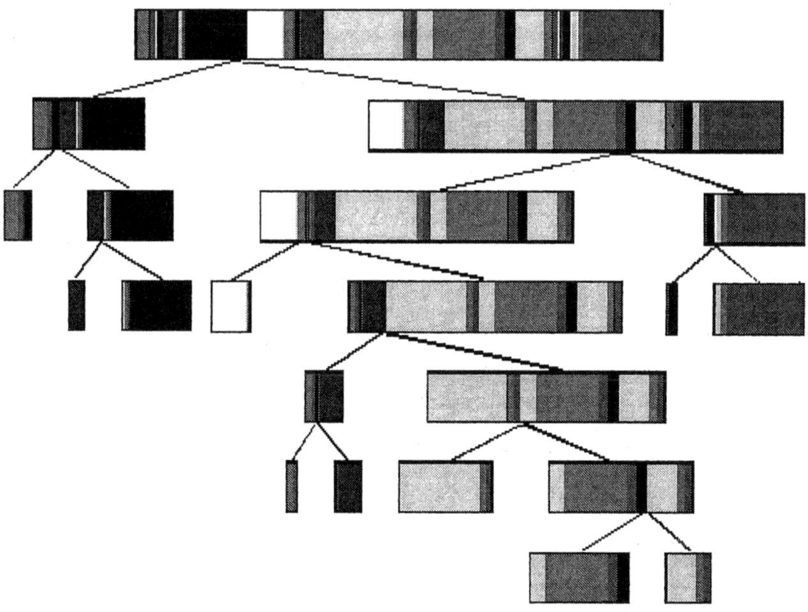

Fig. 1. Tree obtained by referring to the ΔH_O criterion

In particular, for simplicity, we considered trees with 11 and 12 terminal nodes respectively, characterised by misclassification

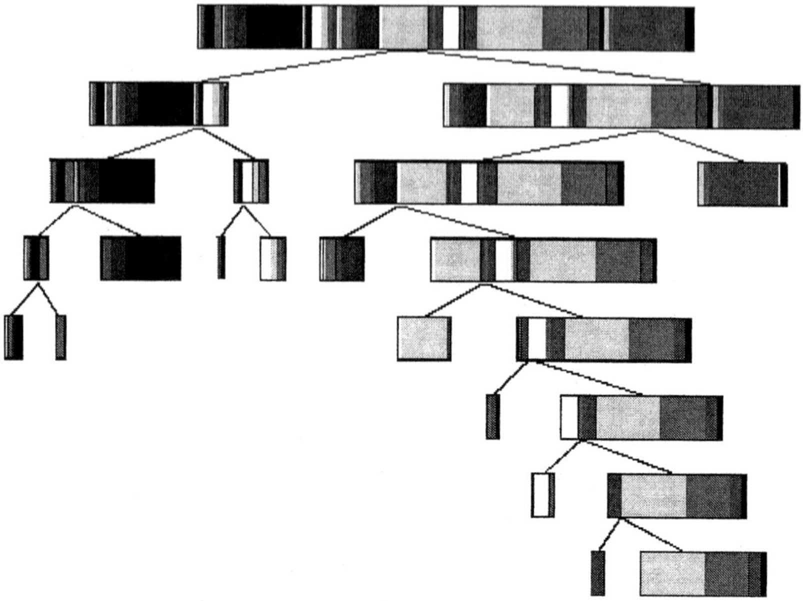

Fig. 2. Tree obtained by referring to the ΔS_O criterion

rates equal to 21.77% and 33.92%. As concerns the tree obtained using ΔH_O (Figure 1), it can be noticed that the first split separates the first 4 classes from the others. Moreover, many terminal nodes appear strongly characterised by the presence of one class (eventually combined with adjacent classes). Considering now the tree obtained by applying criterion ΔS_O (Figure 2), we observe that terminal nodes appear more confused than in the previous case, and not-adjacent classes are put into the same terminal node.

To better compare the obtained results it is worthwhile to consider the confusion tables characterising the classifiers (Tables 1 and 2).

Table 1. Joint distribution of $(Y, \hat{Y}(\Delta H_O))$

$Y \backslash \hat{Y}(\Delta H_O)$	1	2	3	4	5	6	
1	26	0	0	1	0	0	**27**
2	1	82 (60)	13 (34)	4	0	2 (3)	**102**
3	2	11 (5)	61 (53)	4	1 (6)	2 (11)	**81**
4	0	10 (5)	13 (12)	59 (61)	2 (4)	3 (5)	**87**
5	0	0	2 (0)	0	28 (18)	0 (12)	**30**
6	0	1 (0)	13 (9)	0 (6)	1 (0)	53 (53)	**68**
	29	**104**	**102**	**68**	**32**	**60**	**395**

Table 2. Joint distribution of $(Y, \hat{Y}(\Delta S_O))$

$Y \backslash \hat{Y}(\Delta S_O)$	1	2	3	4	5	6	
1	21	3	0	0	2	1	**27**
2	6	87	0	4	2	3	**102**
3	7	36	17	7	8	6	**81**
4	3	12	0	62	4	6	**87**
5	0	0	0	0	18	12	**30**
6	0	4	1	7	0	56	**68**
	37	**142**	**18**	**80**	**34**	**84**	**395**

In Table 1 the joint distribution of the response variable, Y, and the predictor $\hat{Y}(\Delta H_O)$ is represented. In particular, we consider either the predictor determined by the tree with 11 terminal nodes (with rate of misclassification 21.77%) and the one obtained by the tree with 6 terminal nodes having a rate of misclassification, 31.392%, more similar to the one characterising the tree with 12 terminal nodes obtained by referring to criterion ΔS_O (we observed that misclassification rates characterising ΔS_O-classifier are greater than those characterising ΔH_O-classifier for every dimension of the tree). From the above considered tables it seems that concordance between the response variable and its predictor is greater when criterion ΔH_O is used. Standard measures of concordance confirm the impression that concordance between Y

and $\hat{Y}(\Delta H_O)$ is stronger than the concordance between Y and $\hat{Y}(\Delta S_O)$ in both the considered cases – 11 or 6 terminal nodes.

Nevertheless, we are in this case mainly interested in emphasizing the structural differences between the two approaches rather than in evidencing the better performances of one of them. An interesting consideration concerns the fact that the ΔS_O-classifier appears to suffer for the absence of the anti-end-cut factor, a factor contrasting the creations of one (almost pure) sub-node containing a small number of individuals and one containing all the other individuals (see Taylor and Silverman (1993) for a detailed discussion). The above considerations encourage the using of our criterion as a possible valid alternative to the criterion solely used to build classification trees in the ordinal case, ΔS_O.

Needless to say, the proposed ΔH_O will possibly suffer of many of the drawbacks which have already been underlined for its nominal counterpart (see, among the others, Taylor and Silverman (1993) and Shih (1999)). Nevertheless, we think it can constitute a good "starting point", which has to be eventually modified taking into account and adapting to the ordinal case the proposals developed to improve the Gini-Simpson criterion.

5 A Property of the New Criterion

In this section we illustrate an interesting property of ΔH_O (characterising also criterion ΔH_{OS}). In particular, we will show that ΔH_O can be related to an index measuring the proportional reduction of heterogeneity of the response Y due to a predictor. This result is not only interesting from an interpretative point of view, but also in that it permits to apply the algorithm FAST introduced by Mola and Siciliano (1997, 1998) to fasten the process of growing the maximal tree. In particular, let X be a generic predictor of Y, assuming I levels; consider a node t, and suppose cases in t are divided, according to the levels of X, into I subgroups. If $H_O(t)$ is the heterogeneity of Y at node t and $H_O(i|t)$ is the heterogeneity of Y in the i-th subgroup, having a proportion $p_t(i)$ of cases, the proportional reduction in heterogeneity of Y

due to predictor X is:

$$\tau_{O(t)}(Y|X) = \frac{H_O(t) - \sum_{i=1}^{I} H_O(i|t)p_t(i)}{H_O(t)} \tag{14}$$

This index (Piccarreta, 2001) measures the strength of association between the (nominal) predictor X and the (ordinal) response variable Y. Consider now a split s_X, defined on the basis of X, generating the two subgroups t_L and t_R. It can be shown that:

$$\tau_{O(t)}(Y|s) = \frac{H_O(t) - p_L H_O(t_L) - p_R H_O(t_R)}{H_O(t)} = \frac{\Delta H_O(t,s)}{H_O(t)} \tag{15}$$

Hence, the split maximising $\Delta H_O(t,s)$ is the split maximising $\tau_{O(t)}(Y|s)$. This result emphasises that our criterion, as the Gini-Simpson one (see Mola and Siciliano (1997)), searches for the splitting variable predicting the response variable in the best possible way (according to the predictability index τ_O). Moreover, it can be easily shown that $\tau_{O(t)}(Y|X) \geq (\tau_{O(t)}(Y|s_X)$ for each split s_X based upon the levels of X. Following Mola and Siciliano (1997), this property can be used to find the best split of a node without trying out all possible splits.

Consider in fact two explanatory variables, say X and Z, and suppose it is $\tau_{O(t)}(Y|X) \geq \tau_{O(t)}(Y|Z)$. Let s_X^* and s_Z^* be the best splitting splits for node t which can be defined on the basis of X and Z respectively. If $\tau_{O(t)}(Y|s_X^*) \geq \tau_{O(t)}(Y|Z)$, then necessarily $\tau_{O(t)}(Y|s_X^*) \geq \tau_{O(t)}(Y|s_Z^*)$, so that it is not necessary to evaluate the predictive power of all the splits which can be defined on the basis of Z. The computational gain in terms of the reduced number of splits inspected and in terms of CPU time is illustrated in Mola and Siciliano (1997, 1998).

6 Conclusions

In this work we introduce a new criterion for generating classification trees in the case when the levels of the categorical response variable, Y, can be ordered. This is done by referring to a general measure of the mutual dispersion of a variable. Such a measure, introduced by Gini (1954), can be applied to every variable, whatever its nature, by suitably defining a measure of the dissimilarity between a generic couple of values (or categories) assumed by the variable itself. We illustrate that two "natural" choices for the dissimilarity between the values of nominal and quantitative variables lead to the measures to evaluate a split which are at the basis of CART, one of the most popular procedures to build classification trees. We thus define an ordinal criterion by referring to a suitable measure of dissimilarity between the levels of an ordinal categorical variable. The new criterion is compared with the one usually referred to in the ordinal case, either from a theoretical point of view and with respect to their performance, referring to a real data set. The performance of the new algorithm are encouraging, evidencing the possibility to refer to our as a valid alternative to the solely used one.

The main focus of analysis is the *growing phase* of CART procedure, whose final output is the maximal tree, T_M. With respect to this point, we show that our criterion is related to an index measuring the proportional reduction of the heterogeneity of the response variable in a node when splitting the node itself into two sub-nodes. This finding is surely appreciable from an interpretative point of view; moreover, it permits to extend to the ordinal case an algorithm introduced by Mola and Siciliano (1997, 1998) to fasten the procedure of growing T_M.

Of course, the proposed criterion is possibly characterised by some of the drawbacks still evidenced especially for its nominal counterpart. Our aim is to provide a starting point, which should be eventually modified taking into account and adapting to the ordinal case proposals developed to improve the Gini-Simpson criterion.

References

BREIMAN, L., FRIEDMAN, J.H., OLSHEN R.A., and STONE, C.J. (1984): *Classification and Regression Trees.* Wadsworth, Belmont CA.

FABBRIS, L. (1997): *Statistica Multivariata. Analisi Esplorativa dei dati.* Mc-Graw(Hill, Milano.

GINI, C. (1954): *Variabilità e concentrazione.* Veschi, Roma.
MOLA, F. and SICILIANO, R. (1997): A Fast Splitting Procedure for Classification Trees, *Statistics and Computing*, *7*, 209-216.
MOLA, F. and SICILIANO, R. (1998): A General Splitting Criterion for Classification Trees, *Metron*, *56*, 156-171.
PICCARRETA, R. (2001): A New Measure of Nominal-Ordinal Association, *Journal of Applied Statistics*, *28*, 107-120.
SHIH, Y.-S. (1999): Families of Splitting Criteria for Classification Trees, *Statistics and Computing*, *9*, 309-315.
TAYLOR, P.C. and SILVERMAN, B.W. (1993): Block Diagrams and Splitting Criteria for Classification Trees, *Statistics and Computing*, *3*, 147-161.

Part II

Time Series Analysis and Related Methods

Space Time Noisy Observation Smoothing

Di Giacinto Valter[1], Ippoliti Luigi[2], and Romagnoli Luca[3]

[1] Bank of Italy - L'Aquila - digiacinto.valter@insedia.interbusiness.it
[2] Dipartimento di Metodi Quantitativi e Teoria Economica, Facoltà di Scienze Manageriali dell'Università "G. d'Annunzio" di Chieti, Viale Pindaro, 65127 Pescara, Italy - ippoliti@dmqte.unich.it
[3] Istituto Nazionale di Statistica, Ufficio regionale per il Molise, Via Mazzini 129, 86100 Campobasso, Italy - luromagn@istat.it

Abstract. The paper proposes an *adjusted* maximum likelihood estimator for the parametric estimate of a *STARG*$(p, \lambda_0, \cdots, \lambda_p)$ model with measurement noise. Provided the noise variance is known or can be estimated consistently, the *adjusted* maximum likelihood estimator is proved to be asymptotically equivalent to the corresponding *exact* maximum likelihood estimator that, in this study, turns out to be computationally untractable. The theoretic background outlined in the paper finds a natural field of application in observed image sequences. Thus, we present the results of a state-space smoothing procedure performed on monthly observations over a regular lattice.

1 Introduction

The paper is concerned with parameter estimation and smoothing of a spatio-temporal series corrupted by noise. In particular, we assume that $X(\mathbf{s}, t)$ is a spatio-temporal process observed over time t and general location $\mathbf{s}$ within the geographical domain of interest D. That is, we consider $X(\mathbf{s}, t)$ at $(\mathbf{s}_i, t)$ for $i = 1, 2, ..., n$ and $t = 1, 2, ..., T$. In what follows, we use the model

$$X(\mathbf{s}, t) = Y(\mathbf{s}, t) + e(\mathbf{s}, t) \tag{1}$$

where $Y(\mathbf{s}, t)$ is a zero mean, L_2-continuous (i.e. $E\{Y(\mathbf{s} + \mathbf{h}, t) - Y(\mathbf{s}, t)\}^2 \to 0$ as $\mathbf{h} \to \infty$; Cressie 1993, p. 127), second order Gaussian stationary process; $e(\mathbf{s}, t)$ is a white noise measurement process, independent of $Y(\mathbf{s}, t)$, with variance σ_e^2. The variable $Y(\mathbf{s}, t)$ represents the variable of interest and constitutes the state or the signal process. The paper is organised as follows: in Section 2 the state process is assumed to be represented by a $STARG(p, \lambda_0, \cdots, \lambda_p)$ (Space Time Autoregressive Generalised) model (Di Giacinto 1994; Terzi 1995) and expression (1) will be noted as a *STARG+Noise* model. Given the assumptions for the state and measurement error processes, Section 3 deals with estimation of model parameters. Under normality assumptions for the signal and noise processes, maximum likelihood estimators are derived. However the likelihood function turns out to be computationally untractable even for samples of moderate dimension. To overcome this problem, following Dryden *et al.* (2002), we propose an adjusted likelihood function. The adjusted maximum likelihood estimator (*AMLE-ST*) is proved to be asymptotically equivalent to the maximum likelihood (ML) estimators, provided the noise variance is known or can be estimated consistently. Apart from the computational feasibility, *AMLE-ST* have the appeal of not requiring the specification of the entire distribution function of the noise process, but only of the first two moments. In Section 3.1 a simulation study is carried out to asses the performance of the estimator. In this case we have selected some example situations to give a flavour of the types of behaviour that our estimator exhibit under particular conditions. Section 4 shows an application to image sequence analysis. The data set is provided by the Goddard Distributed Archive Center (GDAC), and consists of Earth images derived through Normalized Difference Vegetation Index (NDVI) to study the vegetation dynamics around the globe. At last, Section 5 closes the paper with some final considerations on *AMLE-ST*.

2 The STARG+Noise Model

Let us assume that the unobserved state processes $Y(\mathbf{s}, t)$, apart from a deterministic mean component $\mu(\mathbf{s}, t)$, evolves in a Markov manner so that only recent observation values influence the current time value. The large scale variation, measured by the mean component $\mu(\mathbf{s}, t)$, is assumed to be represented by a parametric spatial trend $f(\mathbf{s}; \beta_t)$ of general functional form and possibly time varying parameters β_t. Under this assumption, using vector notation, we model the spatio-temporal dependence by means of the

$STARG(p, \lambda_0, \cdots, \lambda_p)$ (Space Time Autoregressive Generalised) model (Di Giacinto 1994; Terzi 1995)

$$\mathbf{Y}(\cdot, t) = \sum_{h=0}^{p} \sum_{k=0}^{\lambda_h} \phi_{hk} \mathbf{W}^{(k)} \mathbf{Y}(\cdot, t-h) + \mathbf{u}(\cdot, t) \tag{2}$$

where $\mathbf{Y}(\cdot, t)$ and $\mathbf{u}(\cdot, t)$ are $(n \times 1)$ vectors corresponding to the n sites, p is the temporal order of the model, λ_h is the spatial order of the $h-th$ autoregressive component, ϕ_{hk} is the auto-regressive parameter at temporal lag h and spatial lag k ($\phi_{00} = 0$), $\mathbf{W}^{(\cdot)}$ is the $(n \times n)$ spatial contiguity matrix (Pfeifer and Deutsch, 1980) and $\mathbf{u}(\cdot, t)$ is a homoskedastic white-noise term with variance $\sigma_u^2 \mathbf{I}$. Note that if isotropy is not reasonable, an anisotropic process can be used allowing the coefficient ϕ_{hk} varying also with direction and orientation. Setting

$$\mathbf{A}_0 = \mathbf{I} - \sum_{k=0}^{\lambda_0} \phi_{0k} \mathbf{W}^{(k)}$$

$$\mathbf{A}_h = \sum_{k=0}^{\lambda_h} \phi_{hk} \mathbf{W}^{(k)}, \quad \textit{for } h = 1, 2, ..., p$$

we rewrite the $STARG$ model in its vector autoregressive (VAR) form

$$\mathbf{A}_0 \mathbf{Y}(\cdot, t) - \mathbf{A}_1 \mathbf{Y}(\cdot, t-1) - ... - \mathbf{A}_p \mathbf{Y}(\cdot, t-p) = \mathbf{u}(\cdot, t). \tag{3}$$

to obtain the state-space representation of the *STARG-Noise* model (Ippoliti *et al.*, 1998). In fact, model (3) can also be represented in a more compact notation through the following expressions

$$\xi(\cdot, t) = \mathbf{\Phi} \xi(\cdot, t-1) + \mathbf{U}(\cdot, t) \tag{4}$$

where $\xi(\cdot, t)$ is the *state vector*, $\mathbf{\Phi}$ is the *transition matrix* and $\mathbf{U}(\cdot, t)$ is *the model noise*; their structure is as follows

$$\xi(\cdot,t) = \begin{bmatrix} \mathbf{Y}(\cdot,t) \\ \mathbf{Y}(\cdot,t-1) \\ \vdots \\ \mathbf{Y}(\cdot,t-p+1) \end{bmatrix}$$

$$\mathbf{\Phi} = \begin{bmatrix} \mathbf{A}_1 & \mathbf{A}_2 & \cdots & \mathbf{A}_{p-1} & \mathbf{A}_p \\ \mathbf{I} & \mathbf{0} & \cdots & \mathbf{0} & \mathbf{0} \\ \vdots & \vdots & \cdots & \vdots & \vdots \\ \mathbf{0} & \mathbf{0} & \cdots & \mathbf{I} & \mathbf{0} \end{bmatrix}$$

$$\xi(\cdot,t-1) = \begin{bmatrix} \mathbf{Y}(\cdot,t-1) \\ \mathbf{Y}(\cdot,t-2) \\ \vdots \\ \mathbf{Y}(\cdot,t-p) \end{bmatrix}$$

$$\mathbf{U}(\cdot,t) = \begin{bmatrix} \mathbf{u}(\cdot,t) \\ \mathbf{0} \\ \vdots \\ \mathbf{0} \end{bmatrix}$$

Equation (4) is the state equation of the state-space form of model (3) while the corresponding measurement equation is given by

$$\mathbf{X}(\cdot,t) = \mathbf{H}\xi(\cdot,t) + \mathbf{e}(\cdot,t) \tag{5}$$

where $\mathbf{H} = [\mathbf{A}_0^{-1}\ \mathbf{0} \cdots \mathbf{0}]$ is the *measurement matrix.*

Given the state space formulation, it is straightforward to apply the Kalman filter which, computing the optimal recursive estimation of the state vector at time t, leads to the implementation of algorithms for filtering and smoothing.

3 Maximum Likelihood Estimation

The Kalman filter can be performed only when all the parameters of the state space model are known. To this purpose, we present here a maximum likelihood estimator of such parameters. If the initial values of the process are equal to their unconditional mean, the *STARG* model can also be written as follows

$$\Pi(\tilde{\mathbf{Y}} - \tilde{\mu}) = \tilde{\mathbf{u}} \tag{6}$$

with $\Pi = (\mathbf{I}_T \otimes \mathbf{A}_0) - \sum_{h=1}^{p} (\mathcal{B}_h \otimes \mathbf{A}_h)$ (where the $T \times T$ matrix $\mathcal{B}_h$ has all zero entries except for the $h - th$ lower diagonal, with elements equal to one), $\tilde{\mathbf{Y}}{=}vec(\mathbf{Y})$, $\mathbf{Y} = \{\mathbf{Y}(\cdot, t),\ t = 1, \cdots, T\}$, $\tilde{\mathbf{u}}{=}vec(\mathbf{u})$, $\mathbf{u} = \{\mathbf{u}(\cdot, t),\ t = 1, \cdots, T\}$. Assuming a polynomial approximation for the functional form $f(\mathbf{s}; \beta_t)$ of the deterministic spatial trend, we can set

$$\mu(\cdot, t) = \mathbf{D}\beta_t \tag{7}$$

where β_t is an r-dimensional vector of coefficients and $\mathbf{D}$ is an $n \times r$ suitable design matrix. As a consequence we have

$$\tilde{\mu} = \tilde{D}\tilde{\beta} \tag{8}$$

where $\tilde{D} = (\mathbf{I}_T \otimes \mathbf{D})$ and $\tilde{\beta}{=}vec(\beta)$, $\beta = \{\beta_t,\ t = 1, \cdots, T\}$. When the trend coefficients are fixed over time, i.e. when $\beta_1 = \beta_2 = ... = \beta_T = \beta$, expression (8) simplifies to $\tilde{\mu} = \tilde{D}\beta$, where now $\tilde{D} = (\iota_t \otimes \mathbf{D})$.

In this case, the Gaussian log-likelihood function of the STARG process has a remarkably simple expression and can be directly obtained as

$$L_Y(\mathbf{Y}; \theta | \mathbf{Y}_0) \cong -\frac{nT}{2} \log(\sigma_u^2) + T \log|\mathbf{A}_0| - \frac{(\tilde{\mathbf{Y}} - \tilde{\mu})^T \Pi^T \Pi (\tilde{\mathbf{Y}} - \tilde{\mu})}{2\sigma_u^2} \tag{9}$$

where $\mathbf{Y}_0 = [\mathbf{Y}(\cdot, 0), \mathbf{Y}(\cdot, -1), \cdots, \mathbf{Y}(\cdot, -p+1)]$ and θ is the vector of all the parameters to be estimated. However, since the state process is not observable, using (1) and following Dryden *et al.*

(2002), it is possible to work with the log-likelihood of the signal by means of the observed series $\mathbf{X}(\cdot, t)$

$$L_Y(\mathbf{X};\theta|\mathbf{X}_0) \cong$$
$$-\frac{nT}{2}\log(\sigma_u^2) + T\log|\mathbf{A}_0| - \frac{(\tilde{\mathbf{X}} - \tilde{\mathbf{E}} - \tilde{\mu})^T \Pi^T \Pi (\tilde{\mathbf{X}} - \tilde{\mathbf{E}} - \tilde{\mu})}{2\sigma_u^2}$$
$$\cong -\frac{nT}{2}\log(\sigma_u^2) + T\log|\mathbf{A}_0| - \frac{1}{2\sigma_u^2}(\tilde{\mathbf{X}} - \tilde{\mu})^T \Pi^T \Pi (\tilde{\mathbf{X}} - \tilde{\mu}) -$$
$$-\frac{1}{2\sigma_u^2}\tilde{\mathbf{E}}^T \Pi^T \Pi (\tilde{\mathbf{X}} - \tilde{\mu}) - \frac{1}{2\sigma_u^2}(\tilde{\mathbf{E}}^T \Pi^T \Pi \tilde{\mathbf{E}}) \tag{10}$$

where $\tilde{\mathbf{X}} = vec(\mathbf{X})$, $\mathbf{X} = \{\mathbf{X}(\cdot, t),\ t = 1, \cdots, T\}$, $\tilde{\mathbf{E}} = vec(\mathbf{e})$, $\mathbf{e} = \{\mathbf{e}(\cdot, t),\ t = 1, \cdots, T\}$. Since $E\left[\tilde{\mathbf{E}}^T \Pi^T \Pi (\tilde{\mathbf{X}} - \tilde{\mu})\right] = \sigma_e^2 tr(\Pi^T \Pi)$, we obtain the following *Adjusted* Maximum Likelihood Estimator (*AMLE-ST*)

$$L_Y(\cdot|\cdot) \cong -\frac{nT}{2}\log(\sigma_u^2) + T\log|\mathbf{A}_0| - \frac{\left[\bar{\mathbf{X}}^T \Pi^T \Pi \bar{\mathbf{X}} - 2\sigma_e^2 tr(\Pi^T \Pi)\right]}{2\sigma_u^2} \tag{11}$$

where $\bar{\mathbf{X}} = (\tilde{\mathbf{X}} - \tilde{\mu})$ and tr the *trace* operator. Note that the estimator is weakly consistent and asymptotically normal (see, Dryden *et al.*, 2002). From the first order conditions, a closed form solution can be obtained for the trend coefficients and innovation variance σ_u^2

$$\hat{\tilde{\beta}} = (\tilde{D}^T \Pi^T \Pi \tilde{D})^{-1} \tilde{D}^T \Pi^T \Pi \tilde{\mathbf{x}} \tag{12}$$

$$\widehat{\sigma}_u^2 = \frac{1}{nT}\left[(\tilde{\mathbf{X}} - \tilde{\mu})^T \Pi^T \Pi (\tilde{\mathbf{X}} - \tilde{\mu}) - 2\sigma_e^2 tr(\Pi^T \Pi)\right] \tag{13}$$

and to facilitate the iterative search needed to compute the ML estimates of the autoregressive coefficients, a concentrated log-likelihood function can be derived substituting (13) in (11)

$$L_Y(\mathbf{X}, \cdot|\cdot) \cong T\log|\mathbf{A}_0| -$$
$$-\frac{nT}{2}\log\left[(\tilde{\mathbf{X}} - \hat{\tilde{\mu}})^T \Pi^T \Pi (\tilde{\mathbf{X}} - \hat{\tilde{\mu}}) - 2\sigma_e^2 tr(\Pi^T \Pi)\right] \tag{14}$$

where $\hat{\tilde{\mu}} = \tilde{\mathbf{D}}\hat{\tilde{\beta}}$. Since $\hat{\tilde{\mu}}$ is a non linear function of the autoregressive coefficients ϕ, maximization of (14) is not an easy task.

However it can be greatly facilitated by the adoption of the following stepwise optimisation procedure:

1. a preliminary estimate $\hat{\tilde{\beta}}_0$ of $\tilde{\beta}$ is computed setting $\Pi = \mathbf{I}_{nT}$ in (12), i.e. taking the OLS estimates of the regression of $\tilde{\mathbf{X}}$ on $\tilde{\mathbf{D}}$; given the deterministic nature of $\tilde{\mathbf{D}}$ the OLS estimators are consistent and thus provide a valid and easily computed starting point;
2. $\hat{\tilde{\mu}}_0 = \tilde{\mathbf{D}}\hat{\tilde{\beta}}_0$ is substituted for $\hat{\tilde{\mu}}$ in (14) and this is maximized through an iterative search algorithm (the Newton-Raphson iterative procedure has been successfully implemented to this purpose) to yield the estimates $\widehat{\phi}_0$ of ϕ;
3. based on the estimated $\widehat{\phi}_0$ the matrix $\widehat{\Pi}_0$ is derived and the GLS estimates $\hat{\tilde{\beta}}_1$ are computed from (12);
4. steps 2 and 3 are iterated until convergence is achieved.

If the measurement noise variance σ_e^2 is assumed to be known a priori, equation (12-14) can be directly maximised to estimate the remaining parameters. On the other hand, if σ_e^2 is not known, a consistent estimator or a measurement nugget effect (Cressie, 1993) can be used.

3.1 Simulation study

In order to verify the performance of the estimator we have selected some example situations. In each simulation 100 samples were generated from a *STARG(1,11)+Noise* model with parameters: $\phi_{01} = 0.5; \phi_{10} = -0.4; \phi_{11} = 0.2$ and $\sigma_u^2 = 1$. To study the behavior of our estimator we performed different simulations varying both n and T. For all the simulations, the noise variance parameter σ_e^2 has been fixed to produce a Signal-to-Noise-Ratio (SNR, i.e. the ratio of the signal variance and the noise variance) equal to 5 dB. Even if several methods can be carried out to estimate this parameter, in each simulation the measurement noise variance σ_e^2 was fixed equal to its true value. The results of this simulation study are shown in Table 1-2 where the standard errors are reported in brackets.

Table 1. The means (and standard errors) of the parameter estimates from 100 simulations of a *STARG(1,11)+Noise* model. The true parameters are $\phi_{01} = 0.5$; $\phi_{10} = -0.4$; $\phi_{11} = 0.2$ and $\sigma_u^2 = 1$.

Times	**30**				**60**			
Lattice Size	ϕ_{01}	ϕ_{10}	ϕ_{11}	σ_u^2	ϕ_{01}	ϕ_{10}	ϕ_{11}	σ_u^2
n=16×16	0.504	-0.404	0.208	0.981	0.504	-0.401	0.204	0.984
	(0.012)	(0.011)	(0.019)	(0.028)	(0.008)	(0.008)	(0.013)	(0.016)
n=32×32	0.503	-0.401	0.205	0.984	0.503	-0.402	0.205	0.986
	(0.006)	(0.005)	(0.010)	(0.010)	(0.004)	(0.004)	(0.007)	(0.008)
n=64×64	0.505	-0.403	0.206	0.984	0.504	-0.401	0.203	0.982
	(0.003)	(0.003)	(0.005)	(0.006)	(0.002)	(0.002)	(0.003)	(0.004)

Table 2. The means (and standard errors) of the parameter estimates from 100 simulations of a *STARG(1,11)+Noise* model. The true parameters are $\phi_{01} = 0.5$; $\phi_{10} = -0.4$; $\phi_{11} = 0.2$ and $\sigma_u^2 = 1$.

Times	**100**			
Lattice Size	ϕ_{01}	ϕ_{10}	ϕ_{11}	σ_u^2
n=16×16	0.504	-0.402	0.205	0.983
	(0.006)	(0.006)	(0.010)	(0.012)
n=32×32	0.504	-0.401	0.202	0.986
	(0.003)	(0.002)	(0.005)	(0.006)
n=64×64	0.504	-0.401	0.203	0.981
	(0.002)	(0.001)	(0.003)	(0.003)

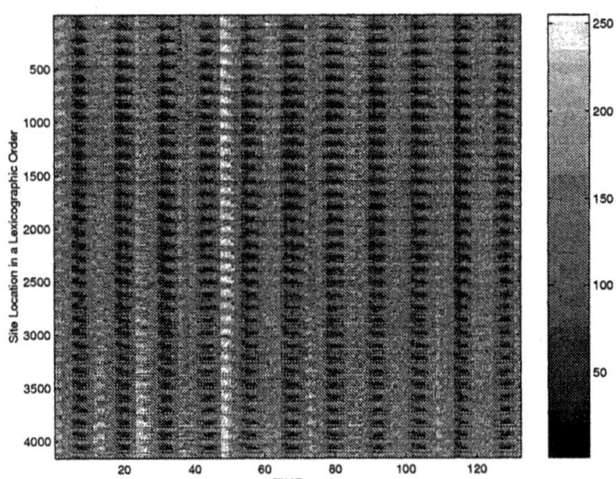

Fig. 1. NDVI data set

As it can be seen from the tables, the fundamental properties of the maximum likelihood estimators are respected. As expected, being a weakly consistent estimator, *AMLE-ST* improves its performance increasing both time and lattice size. Similar finding were obtained in a variety of other simulations with other parameters and spatial neighbourhood structures.

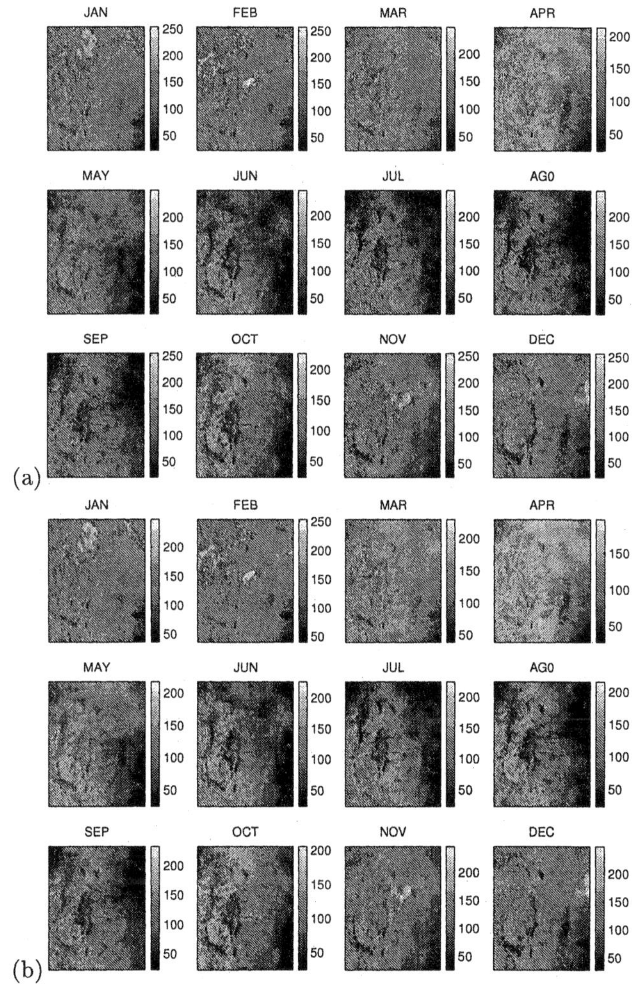

Fig. 2. (a) NDVI data for year 1990; (b) Smoothed NDVI data for year 1990

4 Smoothing of Noisy Image Sequences

The theoretic background outlined in previous sections finds a natural field of application in observed image sequences. Here we present the results of a state-space smoothing procedure performed on monthly observations over a regular lattice.

The data set is provided by the Goddard Distributed Archive Center (GDAC), and consists of Earth images derived through Normalized Difference Vegetation Index (NDVI) to study the vegetation dynamics around the globe. For a more detailed description of the data see (Los *et al.*, 1994). We considered a period of 11 years (January 1982 - December 1992) for a total of 132 temporal observations, each consisting of a 64×64 image from the Middle-East of North America. Figure 1 is a spatial-temporal grey scale image plot of the whole data set where the $N \times M = n$ sites have been arranged in a lexicographic order (i.e. in a raster scan manner). The values range from 0 (for dense vegetation zones) to 255 (for land zones) and the vertical stripes confirm the presence of a temporal seasonality highlighted by the temporal data analysis of the single time series.

To create a zero mean data set we computed a time varying three-parameter plane surface which also allowed for removing the temporal seasonality. Furthermore, although the data were already been treated by the GDAC a measurement noise can still be observed. The estimation of the measurement noise variance σ_e^2 can be carried out through several methods. Huang and Cressie (2000) estimate σ_e^2 as a nugget effect in wavelet decomposition while Gauch (1982) uses eigenvector ordination techniques for field data. Olsen (1993) provides a wide range of methods on this topic. With respect to a single image, reasonables values for σ_ϵ^2 can also be obtained by computing the variance of a flat portion of the image. In this study, following Lim (1990), we employed a pixel-wise adaptive Wiener filtering based on a 5×5 local neighborhood of each pixel obtaining an estimate $\widehat{\sigma_e^2} = 76.91$. This procedure can be easily performed by means of the *wiener2.m* function of Matlab Image Toolbox. Note that we obtained a very similar result detecting σ_e^2 as the nugget effect of a spherical covariogram (Cressie,

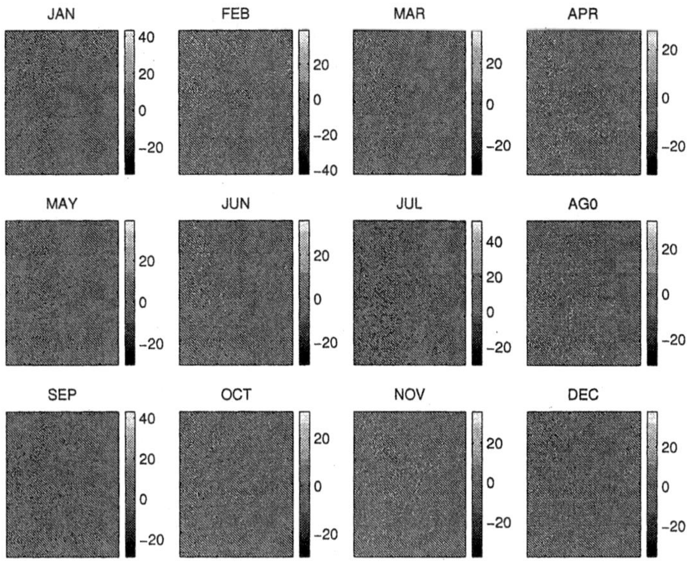

Fig. 3. NDVI estimated measurement error for year 1990

1993). Given the measurement noise variance, the state-space formulation, for different orders of the STARG models, was then used to fit the data and calculate the AIC statistics (Akaike, 1974) to choose the best model. Thus, using this information criteria, the STARG(1,11) model with parameters (standard errors in brackets): $\widehat{\phi_{01}} = 0.809(0.001)$, $\widehat{\phi_{10}} = 0.254(0.001)$,$\widehat{\phi_{11}} = -0.157(0.002)$, $\widehat{\sigma_u^2} = 112.51(0.005)$ was chosen as the best single model as the basis for inference. To give a flavour of the goodness of the smoothing procedure Figures 2(a) and 2(b), present, respectively, the original and filtered data, while Figure 3 shows an estimate of the observational error for the year 1990. To test the goodness of these results it has also been found that smoothing residuals show a covariance pattern represented by a nugget variogram and do not exhibit a spatial structure.

5 Conclusions

Provided a good initial estimate of the noise variance can be given, the adjusted maximum likelihood estimator appears to give excellent results in a variety of situations for noisy *STARG* models. A point to highlight is that *AMLE-ST* depends on the first two moments of the noise distribution, and so would be appropriate for general noise distributions (Dryden *et al.*; 2002). The procedure is reliant on large sample sizes for the consistency arguments in the adjustments to be valid. However, for a sequence of lattice data (such as those considered in Section 4) we usually do have a huge data set. Of course a more deep (theoretic and simulation) study is needed to completely define the statistical property of *AMLE-ST*. To this purpose, this estimator should also be compared with respect to the corresponding exact maximum likelihood estimator (available only for Gaussian measurement noise). All these points, together with the study of the effect of different boundary conditions, will be considered in a future work.

References

AKAIKE, H. (1974) A new look at the statistical model identification. *I.E.E.E. Trans. Auto. Control AC-19*, 716-723

CRESSIE, N. (1993) *Statistics for spatial data.* Wiley, New York.

DI GIACINTO, V. (1994) Su una generalizzazione dei modelli spazio-temporali autoregressivi media mobile (STARMAG). *Atti XXXVII Riunione SIS*, San Remo, aprile 1994, Vol II, 35-42.

DRYDEN, I., IPPOLITI, L. and ROMAGNOLI, L. (2002) Adjusted Maximum Likelihood and Pseudo-Likelihood Estimation for Noisy Gaussian Markov Random Fields. *Journal of Computational and Graphical Statistics, 11(2)*, 370–388.

GAUCH H.G. (1982) Noise Reduction By Eigenvector Ordinations. *Ecology Vol. 63, N. 6*, 1643-1649.

HUANG, H.C., CRESSIE, N. (2000) Deterministic/stochastic wavelet decomposition for recovery of signal from noisy data. *Technometrics 42*, 262-276.

IPPOLITI L., REDFERN E., ROMAGNOLI L.(1998) Kalman Filter on Generalised STARMA Model. *Technical Report*, Department of Statistics, Leeds University.

LIM J.S. (1990) *Two-Dimensional Signal and Image Processing.* Prentice-Hall, London.

LOS, S.O., Justice,C.O., Tucker,C.J. (1994) A global 1 by 1 degree NDVI data set for climate studies derived from the GIMMS continental NDVI data. *International Journal of Remote Sensing, 15(17)*, 3493-3518.

OLSEN, S.I. (1993) Estimation of Noise in Images: an Evaluation. *Graphical Models and Image Processing, 55(4)*, 319–323.

PFEIFER, P. E. and S. J. DEUTSCH (1980) A three-stage iterative procedure for space-time modeling. *Technometrics, 22*, 35-47.

TERZI, S. (1995) Maximum Likelihood estimation of a generalised STAR(p,lp) model. *Journal of the Italian Statistical Society, Vol.4, No.3*

Spectral Analysis in Frequency and Time Domain for Noisy Time Series

Lara Fontanella[1] and Mariagrazia Granturco[1]

Dipartimento di Metodi Quantitativi e Teoria Economica,
Università "G. D'Annunzio" di Chieti-Pescara,
Viale Pindaro, 42 65127 Pescara, Italy

Abstract. In this paper the orthogonal decomposition is used in order to reconstruct the noiseless component of a temporal stochastic process. For weakly stationary processes, the proposed methodology is based on the joint application of the spectral analysis in the frequency domain (Fourier analysis) and in the time domain (Karhunen Loéve expansion). For non stationary processes the orthogonal decomposition is realized in the wavelet domain.

1 Time Series Decomposition and Stochastic Process Expansion

The purpose of this paper is to discuss issues associated with the problem of signal extraction. Using the general orthogonal decomposition of a stochastic process, the object is to analyze the structure of a signal from a set of observations, which occur with error. In particular, assuming that $\{X(t) : t \in [0,T]\}$ is a stochastic process with continuous parameter in the interval $[0,T]$, the measurement equation is given as the sum of two independent components

$$X(t) = L(t) + A(t) \tag{1}$$

where $L(t)$ is the signal and $A(t)$ the noise.
If $X(t)$ is a finite energy process, it is possible to expand it in a set of deterministic functions $\phi_n, n \in \aleph$, which forms a complete orthogonal system in the space of square integrable functions $L_2(T)$

$$X(t) = \sum_{n \in \aleph} a_n \phi_n(t) \tag{2}$$

where the stochastic coefficients a_n are given by the following inner product

$$a_n = \int_T X(t)\phi_n(t)dt \quad n \in \aleph \tag{3}$$

The paper is organized as follows. In Section 2 second order stationary processes are considered and equations (2) and (3) are specified in frequency and time domains. As explained in Section 2.1, we propose a joint application of spectral analysis in these two domains in order to reconstruct the noiseless component $L(t)$ of equation (1). For covariance stationary processes, the proposed methodology also leads to the detection of the harmonic components of the signal. To this purpose, an example, based on the identification of the structural components of the monthly time series of energy consumption in Italy, is presented in Section 2.2. In Section 3, in the context of wavelet analysis, we consider the denoising problem for covariance non-stationary processes. The methodology is illustrated in Section 3.1 while Section 4 concludes the paper with an application to the daily time series of Dioxide Sulphur concentration in the Leeds District.

2 Second Order Stationary Processes: Fourier and Karhunen Loève Expansion

Fourier decomposition, in the frequency domain, and Karhunen Loève expansion, in the time domain, are special cases of equation (2). In the Fourier analysis the basis functions are frequency depending, while in the Karhunen Loève approach they are time depending.

In the Fourier representation the process $X(t)$ is expressed as a special form of a Fourier integral

$$X(t) = \left(\sqrt{2\pi}\right)^{-1} \int_{-\infty}^{\infty} G(w)e^{itw}dw \quad t \in [-T, T] \tag{4}$$

In (4) the coefficients $G(w)$ represent the weights of each basis function e_{itw} in the process reconstruction, and are obtained by the Fourier transform of the process $X(t)$

$$G(w) = \left(\sqrt{2\pi}\right)^{-1} \int_{-\infty}^{\infty} X(t)e^{-itw}dw \tag{5}$$

for which

$$\lim_{T\to\infty} \left\{E\left[|G(w)|^2/2T\right]\right\} = h(w) \tag{6}$$

In (6) $h(w)$ is the non-normalized spectral density function of $X(t)$, obtained as Fourier transform of the covariance function $C(t,s)$

$$\left(\sqrt{2\pi}\right)^{-1} \int_{-\infty}^{\infty} C(t,s)e^{-itw}dt = h(w) \tag{7}$$

On the other hand, in the time domain the Karhunen Loève expansion allows to represent a zero mean random process $X(t)$ as a series of orthogonal functions with uncorrelated coefficients

$$X(t) = \sum_{j=1}^{\infty} z_j\theta_j(t) \quad t \in [0,T] \tag{8}$$

so that

$$z_j = \int_0^T X(t)\theta_j(t)dt \quad j = 1, 2, ... \tag{9}$$

The functions $\theta_j(t)$ are the eigenfunctions of the covariance function of the process and λ_j are the corresponding eigenvalues

$$\int_0^T C(t,s)\theta_j(t)dt = \lambda_j\theta_j(s) \quad j = 1, 2, ... \tag{10}$$

Since the process $X(t)$ is zero mean, the z_j are uncorrelated random variables with mean zero and variance λ_j. Equation (8) is called Karhunen Loève expansion (KLE), though equation (9) is known as Karhunen Loève transform (KLT). As we consider a realization $x(t)$ of the process for n times, $t = 1, ..., n$, a finite approximation for equations (8), (9), (10) is required. In particular equation (10) is now

$$\sum_{t=1}^{n} C(t,s)\widehat{\theta}_j(t) = \widehat{\lambda}_j\widehat{\theta}_j(s) \quad j = 1, 2, \ldots \tag{11}$$

or in matrix form $\mathbf{C}\widehat{\theta}_j = \widehat{\lambda}_j \ \widehat{\theta}_j$ where $\mathbf{C}$ is the covariance matrix of the series. From (11) it follows that the estimators of the eigenfunctions and the eigenvalues of the covariance function are, respectively, the eigenvectors and the eigenroots of matrix $\mathbf{C}$. In the finite approximation KLE and KLT are respectively

$$X(t) = \sum_{j=1}^{n} \widehat{z}_j\widehat{\theta}_j(t) \tag{12}$$

$$\widehat{z}_j = \sum_{t=1}^{n} \widehat{\theta}_j(t)X(t) \quad j = 1, 2, \ldots, n \tag{13}$$

The KLE has the property that, if the eigenfunctions are arranged according to non-increasing order of the eigenvalues, the approximation based on the first p terms minimizes the mean square error (Basilevsky, 1994).

2.1 Detection of Structural Components

For a weakly stationary process, the denoising problem could be solved by means of a linear approximation in the time domain (Mallat, 1998). It is known that the eigenvectors corresponding to the greatest eigenvalues contain information about the temporal correlation and, hence, on the signal $L(t)$. On the other hand, the eigenvectors associated to the smallest eigenvalues are related to the noise component $A(t)$ (Mardia *et al.*, 1998). Several techniques have been proposed to choose the number of eigenvectors to be considered in the reconstruction formula (12) and most of them are described in Mardia *et al.* (1994). In this paper the estimation of the signal is based on the preliminary identification and reconstruction of its additive components. In fact, the stationary signal process can always be decomposed into elementary harmonic waves, which might be aggregated into structural components, as summarized in equation (14)

$$L(t) = C(t) + S(t) \tag{14}$$

where $C(t)$ is a cycle with variable amplitude and $S(t)$ is a seasonal component (Harvey, 1989).

In order to identify and estimate the structural components of a second order stationary process $X(t)$, we propose the joint employment of spectral analysis and Karhunen Loève expansion. The joint application of the above methodologies requires the use of periodogram analysis as a tool to identify the structural components and to test their reconstruction, obtained by the Karhunen Loève technique. Given a time series $x(t)$, in order to obtain a zero mean stationary series we need to remove the trend component, if it exists. The detection of the presence of the remaining structural components in the detrended series $y(t)$ is realized by means of the periodogram. The peaks of the periodogram show the presence of cyclical components with different frequencies. The significance of these peaks is checked by Whittle's and Hartley's tests (Priestley, 1981). The reconstruction of the structural components of the series is obtained by KLE. In (12), with respect to each harmonic component, shown by the periodogram, a linear combination of a subset of eigenvectors of the covariance matrix must be considered. The eigenvectors $\widehat{\theta}_j(t)$ with similar temporal pattern are strongly related to the same basic constituent and then may be aggregated in (12). According to Basilevsky (1994), these eigenvectors correspond to insignificantly different eigenvalues and can be identified by the following test on the equality of the corresponding r eigenroots

$$\chi^2 = -n \sum_{j=1}^{r} \ln\left(\widehat{\lambda}_j\right) + nr \ln\left(\sum_{j=1}^{r}\left(\widehat{\lambda}_j/r\right)\right) \tag{15}$$

This test has a Chi-square distribution with $\left(\frac{1}{2}r(r+1) - 1\right)$ degrees of freedom. The fit of each structural component is reliable if the estimated spectral density function has just a peak at the frequency of the harmonic component that we attempt to reconstruct. To test the goodness of the signal reconstruction we estimate the spectral density function of the residual component, which must be similar to the spectrum of a white noise.

2.2 Application to the Monthly Energy Consumption in Italy (1978-1995)

The proposed methodology has been applied to the monthly energy consumption in Italy from 1978 to 1995 (*Source*: ENEL). In order to obtain a mean stationary time series, a linear trend has been removed (Figure 1.a). Whittle's and Hartley's tests on the periodogram peaks (Tables 1 and 2) confirm the presence of harmonic components in the series.

Table 1. Whittle's Test on the Periodogram Peaks

Frequencies	Test Statistics	Critical value 1%	Frequencies	Test Statistic	Critical value 1%
0.254	0.298	0.083	1.047	0.263	0.087
3.142	0.386	0.084	0.058	0.335	0.087
2.094	0.287	0.084	2.618	0.299	0.088
1.571	0.308	0.085	0.087	0.118	0.089
0.029	0.231	0.086	0.116	0.161	0.090

Table 2. Hartley's Test on the Periodogram Peaks

Test Statistics	Critical value 1%
2.6446	1.999

The significant frequencies highlight the occurrence of cyclical and seasonal components. The eigenvalues scree plot (Figure 1.b), the Chi square test (15) and the eigenvectors periodogram similarities let the identification of the eigenvectors subsets for the estimation of nine harmonic components (Table 3). As stressed by the peaks of the estimated signal periodogram (Figure 1.e), the reconstructed detrended series (Figure 1.c) preserves the same spectral contents of the original one. The goodness of the reconstruction methodology is supported by the estimated noise periodogram (Figure 1.f), whose behavior is similar to the spectrum of a white noise as confirmed by Whittle and Hartley's tests.

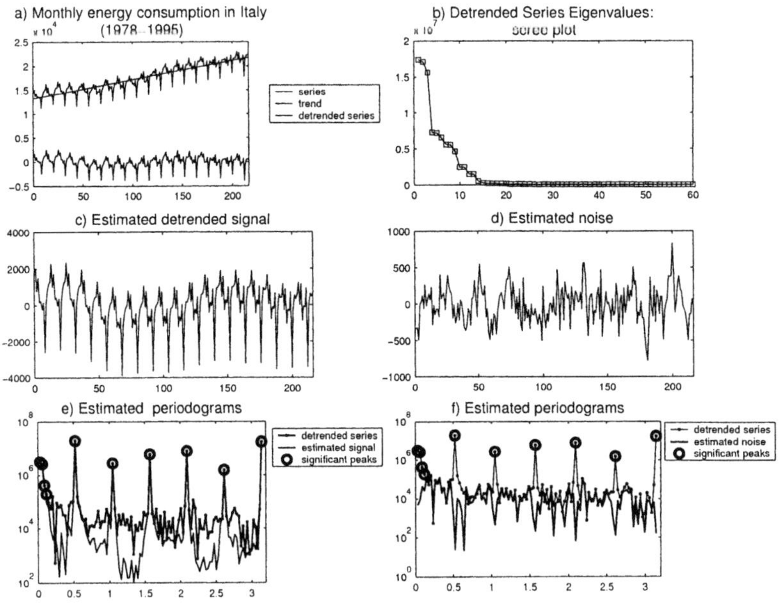

Fig. 1. Monthly energy consumption in Italy (1978-1995).

Table 3 Chi square Test on the eigenvectors

Frequencies	Period in months	Basic components	Eigenvectors subsets	Explained variance %	Test Statistics	Critical value 1%
0.5236	12	$1^{st} : S(t)$	$1^{st} - 2^{nd}$	35	0.019	9.210
3.1416	2	$2^{nd} : S(t)$	3rd	17.9	-	-
2.0944	3	$3^{rd} : S(t)$	$4^{th} - 5_{th}$	17.4	0.005	9.210
1.5708	4	$4^{th} : S(t)$	$7^{th} - 8^{th}$	12.2	0.006	9.210
1.0472	6	$5^{th} : S(t)$	$10^{th} - 11^{th}$	5.3	0.008	9.210
0.0291	216-108	$6^{th} - 7^{th} : C(t)$	$6^{th} - 9^{th}$	4.5	0.001	9.210
2.618	2.5	$8^{th} : S(t)$	$12^{th} - 13^{th}$	2.6	0.010	9.210
0.0873	72	$9^{th} : C(t)$	14th	1.2	-	-

3 Non Stationary Time Processes: Wavelet Expansion

The series expansion (2) of the process in an orthogonal basis can be realized also in the wavelet domain (Mallat 1998). Since the structure of the wavelet basis functions provides specificity in location (via translations) and in frequency (via dilations), this methodology allows a simultaneous analysis of the spectral

contents of the process both in frequency and time domain. The wavelet transform offers a localized frequency decomposition and provides information not only on what frequency components are present on the signal, but also when or where they are occurring. Therefore wavelets have significant advantages over basic Fourier analysis, when the process under study is non-stationary.
In this paper we focus our attention only on the Discrete Wavelet Transform (DWT). In this case the construction of the basis of the Hilbert space $L_2(T)$ is obtained by translating the father wavelet φ (scaling function) and by translating and dilating the mother wavelet ψ (analyzing function). Consequently the orthonormal basis, $\{\varphi_{jo,k}, \psi_{j,k}\}_{j \geq jo, k \in Z}$, for some fixed $j_0 \in Z$, is given by

$$\varphi_{jo,k}(t) = 2^{jo/2}\varphi\left(2^{jo}t - k\right) \tag{16}$$

$$\psi_{j,k}(t) = 2^{j/2}\psi\left(2^{j}t - k\right) \tag{17}$$

In the wavelet basis, equation (2) can be expressed as

$$X(t) = \sum_{k \in Z} c_{jo,k}\varphi_{jo,k}(t) + \sum_{j \geq jo}\sum_{k \in Z} d_{j,k}\psi_{j,k}(t) \tag{18}$$

where the scaling $c_{jo,k}$ and the wavelet $d_{j,k}$ coefficients are given by the following wavelet transforms

$$c_{jo,k} = \int_T X(t)\varphi_{jo,k}(t)dt \tag{19}$$

$$d_{j,k} = \int_T X(t)\psi_{j,k}(t)dt \tag{20}$$

3.1 Detection of the Noiseless Component in Wavelet Domain

For the time-frequency localization property of the wavelet representation, the identification and reconstruction of the signal component of a non stationary time series can be realized in the wavelet domain.
Let $x = \{x(t_1), x(t_2), , x(t_n)\}$ be a time series of length $n = 2^J$,

equally spaced in the time interval $[0, 1]$. The coefficients obtained by the wavelet transforms (19) and (20) allow for representing the time series in a time-frequency domain. For a given level of approximation j_0, the vector $\mathbf{x}$ is wavelet transformed into a new vector $\mathbf{w}$ of the same length. More specifically, given $j_0 = 0$, we can compute one scaling coefficient $c_{0,0}$ and $(2^J - 1)$ detail coefficients $d_{j,k}, j = 0, 1, ..., J-1, k = 0, ..., 2j - 1$ ($\mathbf{w} = (c_{0,0}, \mathbf{d})$).

The transformation from the temporal to the wavelet domain can be expressed in matrix form as $\mathbf{w}$=$\mathbf{Wx}$, where $\mathbf{W}$ is an orthogonal matrix associated to the orthonormal basis chosen. Since the observed data can be considered as a signal $L(t)$, corrupted by a white noise $A(t)$, as in equation (1), the wavelet coefficients $d_{j,k}$ can be also written as the sum of a signal and an erratic component, $d_{j,k} = d^*_{j,k} + \eta_{j,k}$. With the aim to reconstruct the signal, we need to remove the noise component from each coefficient. A technique to achieve free-noise coefficients is wavelet shrinkage (Donoho and Johnstone 1994-1995). Under this scheme "hard" or "soft" thresholded coefficients $\widehat{d^*_{j,k}} = \delta_\Lambda(d_{j,k})$ can be obtained respectively from (21) and (22)

$$\delta_\Lambda(d_{j,k}) = d_{j,k} I(|d_{j,k}| > \Lambda) \tag{21}$$

$$\delta_\Lambda(d_{j,k}) = sign(d_{j,k}) max(0, |d_{j,k}| - \Lambda) \tag{22}$$

Various approaches suggest different choice of the threshold Λ (for a review see Antoniadis, 1995). In this paper, for simplicity, we use the universal threshold $\Lambda_{un} = \sigma_A \sqrt{2logn}$ (Donoho and Johnstone, 1994) where the estimator of the noise standard error σ_A can be the *median absolute deviation* (MAD) of the detail coefficients at the highest level

$$\sigma_A = MAD\{d_{j,k}\} = median\left\{|d_{J-1,k} - median\left\{d_{J-1,k}\right\}|\right\}/0.6745 \tag{23}$$

The inverse wavelet transform (18) applied to the thresholded detail coefficients yields to the reconstruction of the signal component

$$\widehat{L}(t) = c_{0,0}\varphi 0,0(t) + \sum_{j=0}^{J-1}\sum_{k=0}^{2^j-1} \widehat{d}_{j,k}\psi_{j,k}(t) \tag{24}$$

The series components at different frequencies are obtained by applying the reconstruction formula (18) for each level of resolution

$$\widehat{L}_j(t) = \sum_{k=0}^{2^j-1} \widehat{d}_{j,k}\psi_{j,k}(t) \tag{25}$$

The analysis of each basic component can suggest where the different frequencies are localized in time.

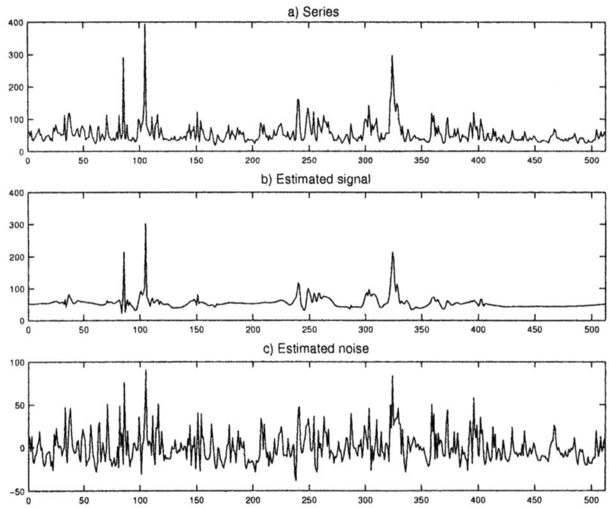

Fig. 2. SO_2 daily mean concentration in the Leeds District (29/03/83-21/08/84): Series, Estimated Signal and Estimated Noise.

4 Application to the Daily Concentration of Dioxide Sulphur (SO_2) in the Leeds District

The wavelet decomposition has been applied to the daily time series of SO_2 mean concentration in the Leeds district (29/03/83-21/08/84) (Mardia et al., 1998).

In the analysis we use a Daubechies with 8 vanishing moments with periodic symmetric boundary condition (Mallat, 1998). The second order non-stationarity of the series (Figure 2.a) is confirmed by the White's heteroskedasticity test (White, 1980). Although the non-stationarity in covariance, the analysis in the wavelet domain allows for separating the signal from the additive noise component. Using hard thresholded coefficients the signal component is estimated by means of (24). In Figure 2.b it is evident that the reconstructed signal is able to track the raw data. The goodness of the signal reconstruction is highlighted by the similarity of the periodograms of the estimated signal and the series (Figure 3.a). Figure 3.b shows the randomness of the estimated noise, confirmed by Whittle's and Hartley's tests on the periodogram peaks.

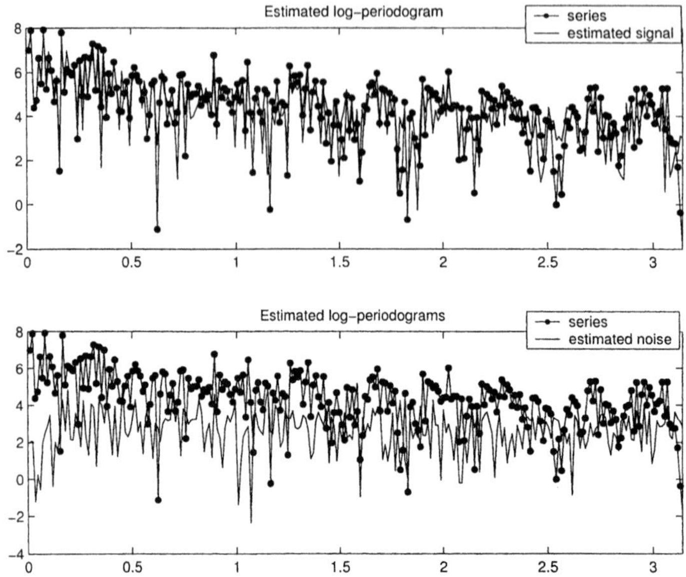

Fig. 3. SO_2 daily mean concentration in the Leeds District (29/03/83-21/08/84): Series, Estimated Signal and Estimated Noise Periodograms.

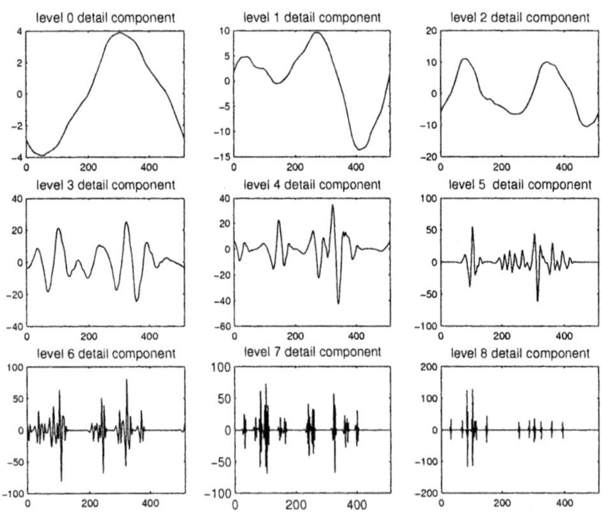

Fig. 4. SO_2 daily mean concentration in the Leeds District (29/03/83-21/08/84): Detail Components at different resolution levels.

The signal reconstruction for each resolution level (Figure 4) emphasizes the wavelet localization property. The lower resolution levels are associated to basic components of small frequencies content and of large spread on the time axis. For higher resolution levels the components of higher frequencies are better localized in time.

References

ANTONIADIS A. (1995): Discussion on Wavelet Shrinkage: Asymptopia. *J.R.S.S., series B, vol 57, No 1.*

BASILEVSKY A. (1994): *Statistical Factor Analysis and Related Methods. Theory and application.* John Wiley and Sons Inc.

DONOHO D. and JOHNSTONE I.(1994):Ideal Spatial Adaptation via Wavelet Shrinkage. *Biometrika, 81*, 425–455.

HARVEY A. C. (1989): Structural Time Series Models and the Kalman Filter. Cambridge Univ. Press.

MALLAT S. (1998): A Wavelet Tour of Signal Processing. Academic Press, London

MARDIA K.V.,KENT J.T. and BIBBY J.M. (1994): Multivariate Analysis, Academic Press, London, 9th Edition.

MARDIA K.V., GOODAL C., REDFERN E.J., and ALONSO F.J.(1998): The Kriged Kalman filter, *TEST, Vol. 7, n2*, 217–285.

PRIESTLEY M.B. (1981): Spectral Analysis and Time Series. Academic Press, London

WHITE H. (1980):A Heteroskedasticity-Consistent Covariance Matrix and a Direct Test for Heteroskedasticity. *Econometrika, 48*, 817–838.

A Resistant Measure of Heteroskedasticity in Explorative Time Series Analysis *

Marcella Niglio[1] and Stefano Maria Pagnotta[2]

[1] Dipartimento di *Scienze Economiche e Statistiche*, Università di Salerno, via Ponte Don Melillo, 84084 Fisciano (SA), Italy
[2] Dipartimento *Persona, Mercato, Istituzioni*, Università del Sannio di Benevento, via Calandra, 82100 Benevento, Italy

Abstract. The stationarity of time series is often reached through the transformation of the observed data. When the analysis of the series is carried out automatically using implemented softwares, it is needed to define some indicators which alerts the system about the non stationarity of the data and leads to right transformations. In this context, the present paper proposes an indicator which detects the heteroskedasticity of the data and its empirical distribution has been investigated through Monte Carlo simulations. The performance of the indicator has been compared to well know homoskedasticity test usually implemented in statistical softwares.

1 Introduction

The seminal work of Box and Jenkins (1976) has given a systematic way to analyze time series. It is based on several steps (specification, estimation, diagnostic checking and forecasting) which often imply complex analysis and requires a deep knowledge of the theoretical background. In order to make the method available to people without expertise, several softwares have been developed to automate the Box and Jenkins procedure and to simplify the analysis of large number of time series. This context includes the TESS System (http://esl.jrc.it/tess/) implemented within the Esprit IV 29.741 project supported by the Eurostat. This software has been recently submitted to an assessment study based on Monte Carlo

* The present paper is partially supported by MURST 2000: *Modelli stocastici e metodi di simulazione per l'analisi di dati dipendenti*. It is a joint work of both authors, anyway Niglio wrote sections 1 and 3, while Pagnotta sections 2 and 4.
Correspondence email: *mniglio@unisa.it*

simulations to evaluate its performance and the ability of the System to extract the trend, the seasonal component and to generate forecasts (Giordano et al., 2000). The analysis of the TESS System starts evaluating the presence of desirable properties in the time series under study such as mean and variance stationarity which have great relevance in model identification and parameters estimation respectively. These properties can be reach carrying out suitable preadjustments of the data which are selected, in a set of different transformations, according to the value of some indicators evaluated from the System. In the present paper a procedure to investigate the heteroskedasticity of the time series is proposed. It is based on two smoothers estimated through the quantiles obtained dividing the data set in subseries of fixed length. An index which alerts the analyst to the presence of heteroskedasticity is derived. A Monte Carlo simulation has been implemented in order to test its performances and to discover its empirical distribution. The proposed procedure is finally compared with the Goldfeld and Quandt test (1965), widely implemented in statistical software for time series analysis. The paper is organized as follows: the theoretical background is shown in section 2; the results of a Monte Carlo experiment and the proposed detection rule are presented in section 3, while some concluding remarks and open problems are in the final section.

2 The Theoretical Background

The classical analysis of time series assumes that the series $X_t, t = 1, \ldots, N$, can be decomposed in four orthogonal unobservable components: the trend T_t, the cycle C_t, the seasonal S_t, and the irregular component U_t. These components can be differently combined providing time series with different features.

In the following we assume that the time series X_t has an additive structure with a constant cycle such that:

$$X_t = T_t + S_t + U_t, \tag{1}$$

where the trend is linear, the seasonal component has constant amplitude and phase and the irregular component is an ARMA(p,q) process. These requirements provides an homoskedastic time series model.

Our aim is to derive a synthetic measure which highlights possible changes in variability of the time series. The construction of this measure is based on two steps: first we consider two resistant smoothers, and then the smoothed values, obtained from each of them, are regressed over time with a simple regression model. The difference η of the estimate of the two slopes is our index of interest. Our conjecture is that when the time series is homoskedastic $\eta \approx 0$, while in presence of heteroskedasticity $\eta \neq 0$ and its value is a function of the overall variability.

Given an integer $k << n$, for each $t = k+1, k+2, \ldots, n-k$ we consider a window:

$$W_t = \{X_{t-k}, X_{t-k+1}, \ldots, X_{t-1}, X_t, X_{t+1}, \ldots, X_{t+k}\} \qquad (2)$$

where the number of elements in W_t is $2k+1$, while the number of all windows is $n-2k$.

For each W_t we compute the smoothers:

$$\begin{aligned} Q_1(t) &= 0.25 \textit{ sample quantile of } W_t \\ Q_3(t) &= 0.75 \textit{ sample quantile of } W_t. \end{aligned} \qquad (3)$$

Let $b^{(1)}$ and $b^{(3)}$ the OLS estimates of the two slopes of the linear regressions:

$$\begin{aligned} Q_1(k+t) &= a^{(1)} + b^{(1)}t + \epsilon_t, \quad t = 1, 2, \ldots, n-2k \\ Q_3(k+t) &= a^{(3)} + b^{(3)}t + \gamma_t, \quad t = 1, 2, \ldots, n-2k. \end{aligned} \qquad (4)$$

The slope $b^{(i)}$ can be written as weighted mean of the smoothed values $Q_i(t)$ by using standard results of simple linear regression models:

$$b^{(i)} = \sum_{t=1}^{n-2k} d_t Q_i(k+t) \text{ and } d_t = \frac{t - \frac{n-2k+1}{2}}{\sum_{i=1}^{n-2k}\left(i - \frac{(n-2k+1)}{2}\right)^2} \quad i = 1, 3 \qquad (5)$$

Our proposed measure to evaluate changes in variability is then

$$\eta = b^{(3)} - b^{(1)} = \sum_{t=1}^{n-2k} d_t(Q_3(k+t) - Q_1(k+t)).$$

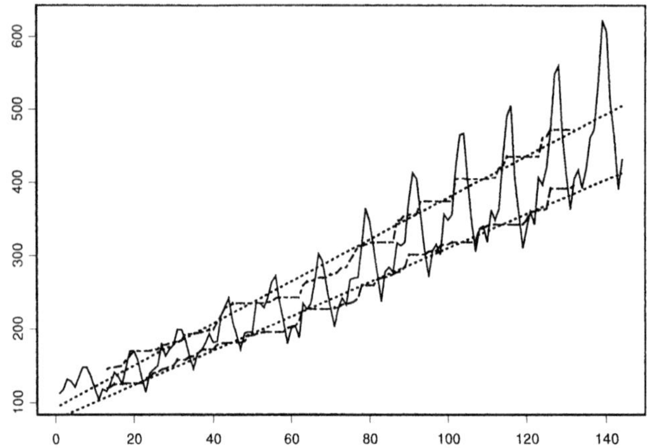

Fig. 1. Airline Passenger time series (solid line) with the resistant smoothers $Q_1(t)$ and $Q_3(t)$ (sketched line) and the lines r_1 and r_3 (dotted line)

We can show, by using standard geometric interpretation on slopes and goniometric formula, that η is proportional to the goniometric tangent of the angle $\Delta\theta = \theta_3 - \theta_1$, where θ_i, $i = 1, 3$ is the angle among the time-axis and the straight lines r_1 and r_2 whose equation are the linear regression models in (4) respectively.

Our conjecture can now be graphically presented. Figures 1 shows the well known Airline passengers time series data (solid line) together with the smoothers (sketched line) and the regression lines (dotted line). The regression lines diverge as the time increases and then the value of η is 0.5306.

In Figure 2 the same time series has been previously *log*-transformed and then the smoothers and the regression lines have been estimated. In this case the regression lines are almost parallel providing a value of $\eta = 0.00$.

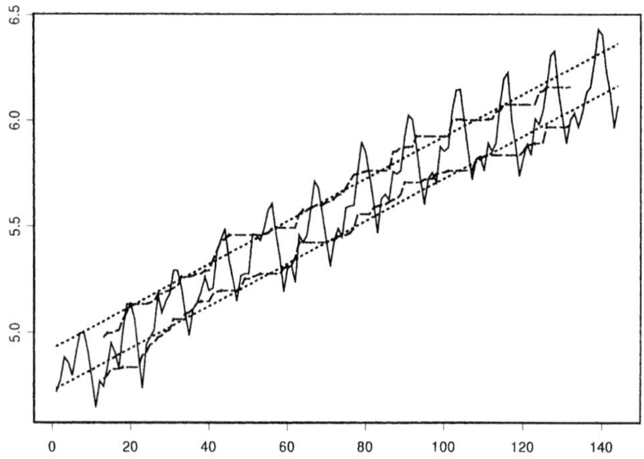

Fig. 2. *Log* of the Airline Passenger time series (solid line) with the resistant smoothers $Q_1(t)$ and $Q_3(t)$ (sketched line) and the lines r_1 and r_3 (dotted line)

The proposition we present assures that under the hypotheses stated above the value of η has to be zero.

Proposition 1. *Let X_t, $t = 1, 2, \ldots, n$ be an homoskedastic time series, following the additive model (1) with orthogonal components; let*

$$\eta = \sum_{t=1}^{n-2k} d_t(Q_3(k+t) - Q_1(k+t)),$$

where Q_i, $i = 1, 3$ is introduced in (3) and d_t is defined in (5).

It follows that

$$\lim_{n\to\infty} E(\eta) = 0$$

Proof.

$$E(\eta) = E\left(\sum_{t=1}^{n-2k} d_t(Q_3(k+t) - Q_1(k+t))\right) =$$
$$= \sum_{t=1}^{n-2k} d_t E(Q_3(k+t) - Q_1(k+t)) = \dots$$

Let $\tau_{i,t}$ be an index such that $Q_i(k+t) \equiv X_{\tau_{i,t}}$, $i = 1, 3$, then

$$\dots = \sum_{t=1}^{n-2k} d_t E(X_{\tau_{3,t}} - X_{\tau_{1,t}}) =$$
$$= \sum_{t=1}^{n-2k} d_t E\left(T_{\tau_{3,t}} + S_{\tau_{3,t}} + U_{\tau_{3,t}} - (T_{\tau_{1,t}} + S_{\tau_{1,t}} + U_{\tau_{1,t}})\right) =$$
$$= \sum_{t=1}^{n-2k} d_t E(T_{\tau_{3,t}} - T_{\tau_{1,t}}) + \sum_{t=1}^{n-2k} d_t E(S_{\tau_{3,t}} - S_{\tau_{1,t}}) +$$
$$+ \sum_{t=1}^{n-2k} d_t E(U_{\tau_{3,t}} - U_{\tau_{1,t}}) = \sum_{t=1}^{n-2k} d_t(T_{\tau_{3,t}} - T_{\tau_{1,t}}) +$$
$$+ \sum_{t=1}^{n-2k} d_t(S_{\tau_{3,t}} - S_{\tau_{1,t}}) + \sum_{t=1}^{n-2k} d_t E(U_{\tau_{3,t}} - U_{\tau_{1,t}});$$

T_t is a continuous function of t then, by using the Lagrange theorem,

$$T_{\tau_{3,t}} - T_{\tau_{1,t}} = c_t(\tau_{3,t} - \tau_{1,t}),$$

for each t. It follows:

$$|T_{\tau_{3,t}} - T_{\tau_{1,t}}| \leq C|\tau_{3,t} - \tau_{1,t}| \leq 2kC$$

with $C = \max_t c_t$ and $2k = \max_t |\tau_{3,t} - \tau_{1,t}|$. Given these inequalities we have:

$$\left|\sum_{t=1}^{n-2k} d_t(T_{\tau_{3,t}} - T_{\tau_{1,t}})\right| \leq \sum_{t=1}^{n-2k} |d_t||(T_{\tau_{3,t}} - T_{\tau_{1,t}})| \leq 2kC \sum_{t=1}^{n-2k} |d_t|$$

where the last summation goes to zero as n goes to infinity, in fact:

$$\sum_{t=1}^{n-2k} |d_t| = \frac{\sum_{t=1}^{n-2k} |t-\bar{t}|}{\sum_{t=1}^{n-2k} |t-\bar{t}|^2} = \\ = \frac{\mathcal{O}(n^2)}{(n-2k)\left[(n-2k)^2-1\right]/12} \sim \mathcal{O}\left(\frac{1}{n}\right) \underset{n\to\infty}{\to} 0.$$

With similar steps also the summation about the seasonal S_t component vanishes as n goes to infinity.

On the other side the summation about the stochastic component U_t is zero because

$$\sum_{t=1}^{n-2k} d_t E(U_{T3,t} - U_{T1,t}) = \sum_{t=1}^{n-2k} d_t \left[E(U_{T3,t}) - E(U_{T1,t})\right] \\ = \sum_{t=1}^{n-2k} d_t(\xi_{0.75} - \xi_{0.25}) = (\xi_{0.75} - \xi_{0.25}) \sum_{t=1}^{n-2k} d_t = 0$$

being $\xi_{0.25}$ and $\xi_{0.75}$ the population quantiles (Rohatgi, 1984, p.616) of the irregular component U_t and $\sum_{t=1}^{n-2k} d_t \equiv 0$.

◇

The sampling distribution of η is investigated through a Monte Carlo simulation discussed in the next section.

3 Empirical Results and Test Procedure

The Monte Carlo study has been carried out considering model (1) where the linear trend is generated as:

$$T_t = 0.04 + 0.008t \tag{6}$$

and the seasonal component is a Fourier transform:

$$S_t = A_0 + \sum_{i=0}^{s/2} [A_i \cos(\omega_i t) + B_i \sin(\omega_i t)] \tag{7}$$

with $A_i = Rcos(0.05)$, $B_i = -Rsin(0.05)$ and $R = 2$. The irregular component is an ARMA(p,q) process:

$$\phi_p(B)U_t = \theta_q(B)\varepsilon_t \tag{8}$$

with gaussian white noise errors, ϵ_t, and coefficients of ϕ and θ shown in Table 1. The autoregressive and moving average orders of the ARMA(p,q) components are p = 1 and $q \in \{1, 2, 3\}$ respectively. The combination of the different orders give rise to 3 additive homoskedastic models, labeled in the following with O11, O12 and O13. For each of them 1000 time series are generated, with seasonal period $s = 12$ and length $n = 200$.

Table 1. Coefficients of the simulated models

Coefficient	ϕ_1	θ_1	θ_2	θ_3
Value	0.57	0.36	0.12	0.11

For all simulated series the η parameter has been estimated and the descriptive statistics of its empirical distribution are shown in Table 2. The mean of the distributions is always zero, according to the theoretical results in the lemma. The variance also appears constant with respect to q, while some differences are in the shape indexes.

Table 2. Descriptive statistics of the sampling distribution of η

Model	Mean	Variance	Skewness	Kurtosis
O11	0.000	9.57 E-6	0.0431	3.1339
O12	0.000	9.26 E-6	0.0312	0.0607
O13	0.000	9.99 E-6	-0.1069	2.6734

The results in Table 2 allow to joint the three different distributions into one. The sampling quantiles of the joint distributions are used to define a double side critical region of size $\alpha = 0.05$, which can be used to test the homoskedasticity of the time series. The empirical rule is:

$$|\hat{\eta}| \geq \eta_0 \tag{9}$$

where $\eta_0 = 0.0061$. It is evaluated as the mean value between $|\hat{F}^{-1}(0.025)|$ and $|\hat{F}^{-1}(0.975)|$, being $\hat{F}^{-1}$ the inverse of the sampling joint distribution function.

In order to assess the performance of (9), heteroskedastic time series have been generated through two more models. The first is an heteroskedastic additive model (E) obtained combining the unobserved components as in the homoskedastic case but with amplitude of the seasonal component dependent from t as:

$$R = 0.03 + 0.01t$$

In this case, the size of the seasonal oscillations increases with the level of the series, as is frequently observed in most economic time series.

Finally, the multiplicative model (M):

$$X_t = T_t \times S_t \times U_t \tag{10}$$

is given by the product of the three components in (6), (7), and (8). This model also generates heterosckedatic time series.

The sampling distributions of η obtained through the additive models, both homoskedastic and heteroskedatic, are shown in Figure 3. The parallel box plots highlight the symmetry and the equal variability of the empirical distributions of η when additive models are examined, even thought the center of the distributions can be different when homoskedastic or heteroskedatic time series are considered.

The distribution of η in the multiplicative case cannot be graphically compared with the previous distributions because of scale inflation. The corresponding plots are in Figure 4 where the distribution of model M11 is remarkably different from models M12 and M13 both in terms of position and variability.

The performance of the empirical rule (9) has been assessed through the comparison between its rejecting percentage with

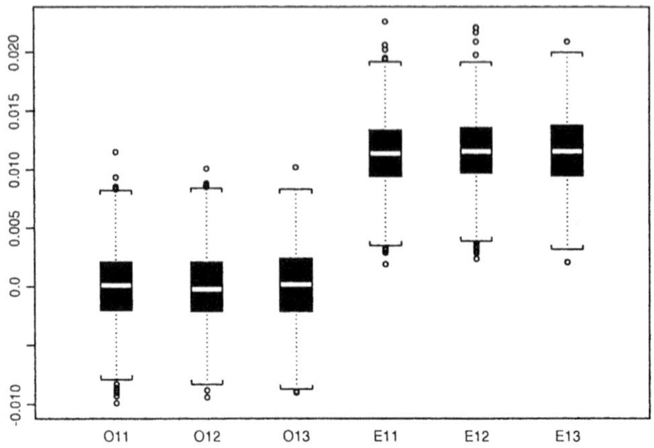

Fig. 3. Parallel box plots of the distribution of η for the homoskedastic and heteroskedastic additive models

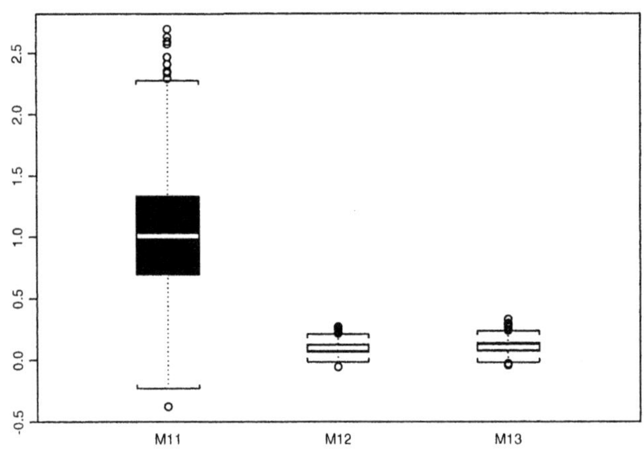

Fig. 4. Parallel box plots of the distribution of η for the multiplicative models

that of the homoskedasticity test of Goldfeld and Quandt (1965). The numerical results are in Table 3.

In the homoskedastic case the proposed rule has an empirical level closer to the nominal one than the G-Q test. When heteroskedastic time series are considered, the results are almost the same. For this experiment the value of k in (2) has been fixed at 12. Further investigation with $k = 24$ does not change the results.

Table 3. Percentage of the simulated time series satisfying the rule (9) and the Goldfeld-Quandt homoskedasticity test, with $\alpha = 0.05$

	O11	O12	O13	E11	E12	E13	M11	M12	M13
Rule (9)	0.049	0.050	0.045	0.960	0.966	0.959	1	0.995	0.990
G-Q test	0.058	0.079	0.066	0.972	0.986	0.984	0.996	0.999	0.995

4 Concluding Remarks

In this paper we have proposed and index η detecting the presence of heteroskedasticity in a time series. This index inherits the property of resistance from the two smoothers needed for the evaluation which are based on the sample quartiles 1 and 3. We explored the theoretical behavior of the index in the case of an homoskedastic time series and we found that the index η has expected value equal to zero in this case. The theoretical convergence to this value is better as the length of the time series increases. The empirical experiments confirms such a results.

Given these results, and taking into account the knowledge about the sampling distribution of η provided by the Monte Carlo experiment, we have also proposed a simple significance test where the null hypothesis states that the time series is homoskedastic. This test have been compared to the one of Goldfeld and Quandt. We found that our proposal has the empirical level of significance closer to the nominal one than in the case of the Goldfeld and Quandt test.

Further investigations, not shown in the paper, have been performed in order to evaluate the robustness of the index η when the time series are contaminated with additive outliers. The encouraging results obtained represent the starting point of the future promising direction of the research.

Another direction has been the extension of the results shown in sections 2 and 3 to the case of more complex trend and seasonal structures.

Acknowledgement: We thank the anonymous referee for his comments and suggestions allowing us to improve this paper.

References

BOX, G.E.P. and COX, D.R. (1964): An analysis of transformations, *Journal of the Royal Statistical Society (B)* , *26*, 211-243.

BOX, G.E.P. and JENKINS, G.M. (1976): *Time series analysis, forecasting and control*, San Francisco: Holden Day.

GIORDANO F., NIGLIO, M. and STORTI, G. (2000): A simulation study for the evaluation of the seasonal adjustment and forecasting performances of the TESS system, *Statistica Applicata*, *12*, 341-360.

GOLDFELD, S.M. and QUANDT, R.E. (1965): Some tests for homoscedasticity, *Journal of the American Statistical Association*, *60*, 539-547.

JARQUE, C.M. and BERA, A.K. (1987): A test for normality of observations and regression residuals, *International Statistical Review*, *55*, 163-172.

ROTHAGI, V.K. (1984): *Statistical inference*, John Wiley & Sons

Part III

Computer Intensive Techniques and Algorithms

Smoothing Score Algorithm for Generalized Additive Models

Claudio Conversano

Dipartimento di Economia e Territorio, Facoltà di Economia, Università di Cassino, Via Mazzaroppi, I-03043, Cassino (FR), Italy

Abstract. In the framework of Generalized Additive Models (GAM) an automatic data-driven procedure is introduced for assigning an appropriate smoother to each covariate and for defining an ordering entrance for the covariates in the model. The resulting *Smoothing Score* algorithm aims to improve model indentifiability. It uses the bagging procedure in order to select the smoothers to be assigned to each covariate and a new scoring measure able to rank the candidate smoothers with respect to their bagged predictive accuracy. The adequacy of this scoring measure is evaluated on artificial data. A comparison between the smoothing score algorithm and the standard GAM is made using real data concerning a classification task.

1 The Framework

Generalized Additive Models (GAM), represents a powerful tool for dealing with nonlinear data structures in the framework of regression analysis. According to Hastie and Tibshirani (1990), the GAM model is defined as:

$$E[Y|\mathbf{X}] = G\left\{\alpha + \sum_{j=1}^{d} f_k(X_j)\right\} \tag{1}$$

In eq. 1 the conditional expectation of the response variable Y is modelled with an additive combination of d predictors, whose influence on the response is evaluated according to the associated smoother, corresponding to one of the possible smoothing functions $f_k(\cdot)$, $(k = 1, \ldots, K)$, in a set of K possible alternative smoothers. These functions, generally known as scatterplot smoothers, ranges from non-parametric tools - such as locally weighted polynomials (known as lowess), smoothing splines, kernel and near-neighbor estimators - to linear parametric regression estimators such as simple and polynomial regression. The estimation procedure is based on the backfitting algorithm, which cycles through the individual terms $f_k(X_j)$ in the additive model and updates them by smoothing suitably defined partial residuals. The link function G can usually be specified as the identity link, used for linear and additive models for Gaussian response data or as a function belonging to the exponential family distribution. In this case, a local scoring algorithm using G produces in each iteration a new working response and a system of weights, which are handed to a weighted backfitting algorithm.

2 GAMs Model Identifiability

Although GAM are very popular among most of the statisticians, they lack with respect to the problem of the correct model identifiability. In fact, by changing the ordering entrance of the predictors in the model as well as the type of their associated smoother (or some smoothing parameter), the result of the final estimation could change considerably. Actually, Hastie (1992) introduced a stepwise selection procedure for GAM based on the Akaike Information Criterion (*AIC*). This procedure, implemented in the S-Plus routine `step.gam`, allows the user to step through arbitrary models in a prespecified path, namely, once that the user expressly defines an *ordered regimen* of candidate smoothers for each predictor to be included in the model. From a practical viewpoint, it is impossible to specify a-priori and arbitrarily the whole set of possible smoothers (and the corresponding degree of smoothness) for each predictor. Some empirical trials demonstrated that this procedure is very sensitive to small perturbations in the data and it stalls when dealing with a lot of predictors.

In the following, in order to find some remedies for these drawbacks, we introduce an automatic procedure for a simultaneous selection of smoothers and predictors in GAM. Following the approach recently introduced by Conversano (2001) for the General Additive Multi-Mixture Models (GAM-MM) of Conversano et. al (2002), the idea is to use both the bagging procedure and a smoother scoring criterion as tools to make a simultaneous and automatic selection of the predictors (with their associated smoothing functions) in the first iteration of the backfitting algorithm, in order to improve the model identification. The outline of the paper is the following: section 2 provides some definitions and an example of bagging-based scoring functions for smoothers and variable selection; in section 3 the estimation procedure based on the definition of an ordering entrance for the predictors in the model is introduced; the benchmarking of our methodology is proposed in section 4 by analyzing a real data set and some concluding remarks are resumed in section 5.

3 Bagging and Scoring for Smoothers Selection

3.1 Notation

We define a set of K alternative smoothers $f_k(\cdot)$, $(k = 1, \ldots, K)$, presenting different degrees of smoothness. Our procedure selects the most appropriate $f_k(\cdot)$ for each predictor to be included in the model. Then, we are dealing with the problem of assigning the appropriate smoother $f_k(\cdot)$ for each candidate predictor X_j, $(j = 1, \ldots, d)$ to be included in the model. On this purpose, for the j-th predictor we define:

- $\phi[f_k(X_j, b)]$: a function ranging in $[0, 1]$ expressing the adequacy of each candidate smoother for the current predictor. Without loss of generality, we assume that $\phi(\cdot)$ is a nondecreasing function, that depends on the data (i.e. the j-th predictor) as well as on the goodness of fit provided by each smoother $f_k(\cdot)$ to the data.
- $v(f_k)$: a function ranging in $[0, 1]$ expressing the cost for the noncomplexity induced by the k-th smoother, depending on its number of parameters.
- $\psi\{[f_k(X_j, b)], v(f_k)\}$: a normalized scoring function depending upon a combination of $\phi(\cdot)$ and $v(\cdot)$.

For the selection of the most appropriate smoother for each predictor X_j, we draw B bootstrap samples from the data. For the b-th bootstrap sample $(b = 1, \ldots, B)$, we search for the smoother presenting the highest score, i.e. the highest value of the scoring function $\psi(\cdot)$. On this purpose, we express the result arising from the b-th sample as follows:

$$\rho_{kbj} = \psi\{\phi[f_k(X_j, b)], v(f_k)\} \tag{2}$$

In eq. 2, ρ_{kbj} is a normalized measure pointing out the smoother assigned to the j-th predictor, such that $\sum_{k=1}^{K} \rho_{kbj} = 1$. For each bootstrap replication, we compute the ρ_{kbj}'s and obtain the most suitable smoother selection for each predictor X_j over all the B bootstrap samples according to a consensus criterion. Practically, the choice of the most appropriate smoother (with the associated degree of smoothness) for each predictor is a synthesis of the scoring functions evaluated in each bootstrap sample.

3.2 The Smoothing Score Function

With respect to eq. 2, possible characterizations of the scoring function could be the Mallows C_p statistics or the Aikaike AIC statistics. When using these measures in order to summarize bootstrap results, some difficulties arise when trying to summarize the results coming out from the different bootstrap replications.
For such reason, we introduce an empirical measure that takes into account, for each combination of candidate smoother and smoothing parameter, the gain in the fit it provides with respect to the most parsimonious alternative, and downgrade this gain with respect to the complexity induced by its number of parameters. This function, that we named *Smoothing Score*, is defined as:

$$\rho_{kbj} = \left(1 - \frac{acc_{kbj}}{acc_{1bj}}\right)\left(1 - \frac{\tau_k}{\tau_K}\right) \tag{3}$$

where $k = 1, \ldots, K$ denotes the smoothing functions, to be ordered from the most to the least parsimonious, associated with the j-th predictor in the *b-th* bootstrap sample.
In coherence with eq. 2 we set $\phi[f_k(X_j, b)] = 1 - acc_{kbj}/acc_{1bj}$, so that acc_{kbj} and acc_{1bj} are measures of accuracy (as for example the Residual Sum of Squares) provided by the k-th smoother and the most parsimonious smoother respectively. Moreover, we set

$v(f_k) = 1 - \tau_k/\tau_K$, so that τ_k and τ_K are the number of parameters of the k-th smoother and the less parsimonious smoother respectively.
In the b-th sample, for each predictor X_j the most appropriate smoother is selected according to the highest value of eq. 3. In this respect, this choice has to be considered optimal because it counts for the trade off between the fitting each smoother provides and the complexity induced by the number of associated parameters.
Therefore, in the b-th sample we use an indicator variable δ_{ljb} such that:

$$\delta_{ljb} = \begin{cases} 1 & \text{if } l = \arg\max_k(\rho_{kjb}) \\ 0 & \text{otherwise} \end{cases} \tag{4}$$

In this way, for each predictor X_j the *bagged smoother selection* is obtained according to the majority vote criterion, here specified as:

$$f_j^* \equiv \arg\max_l \left(\frac{1}{B} \sum_b \delta_{ljb} \right) \tag{5}$$

so that f_j^* denotes the final most voted smoother for the j-th predictor. The proportion of times this smoother has been selected indicates the degree of reliability for the relation between the predictor and the response variable.

3.3 Computational Issues

We have implemented the proposed procedure for smoother and variable selection in GAM in an S-Plus routine, that we named `bag.smooth.score`. Here, a complete set of candidate smoothers is a-priori built in, and the user provides the set of smoothing parameters in the `span` argument. The routine has the following syntax:

```
bag.smooth.score(data,span,index,boot)
```

It requires as input arguments the name of the data frame (`data`), the vector of possible smoothing parameters (`span`), the `index` evaluating the fit produced by each scatterplot smoother and the number of bootstrap replications (`boot`) for producing a bagging evaluation of this fit. Once that a sample has been drawn from the original data set, the `bag.smooth.score` function computes the fitting provided by each scatterplot smoother with respect to a user specified vector of possible smoothing parameters, and for each predictor it chooses the smoother according to a suitable measure. Once that the resampling scheme is fulfilled, it returns for each predictor the *most-voted* smoother f_j according to eq. 5.

3.4 An Example on Simulated Data

In this section, we present the results of a simulation study in order to show how the Smoothing Score function works in practice. Starting form a vector X of $n = 100$ independent and uniformly distributed observations we generated two different data structures, termed as `sim1` and `sim2`, according to the following equations:

$$\begin{array}{lll} \text{sim1:} & y_i = 5 + 1.6x_{i1} + \varepsilon_i & \varepsilon \sim N(0, 0.25) \\ \text{sim2:} & y_i = 5 - 4\sin(\pi x_{i1}) + log(\pi x_{i1}) + \varepsilon_i & \varepsilon \sim N(0, 2.5) \end{array}$$

with $i = 1 \ldots, n$. We obtained two different structures, of the linear type (first case) and of nonlinear type (second case). Scatterplots of the two data frames are showed in fig. 1. For these

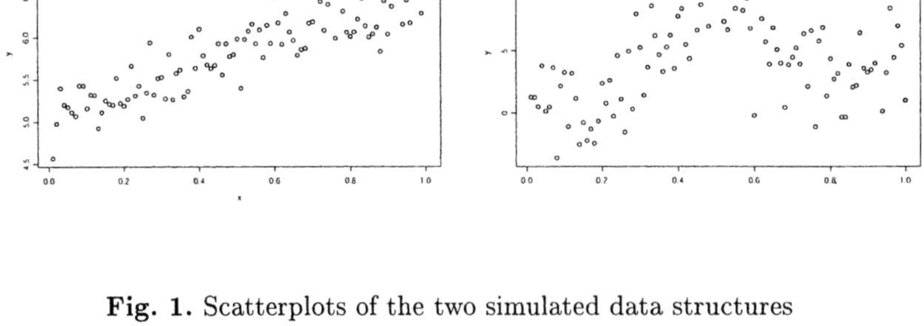

Fig. 1. Scatterplots of the two simulated data structures

structures, we evaluated the outcome of eq. 2 (with respect to

$b = 100$ bootstrap replications) and the behavior of the Smoothing Scores function $\rho(\cdot)$ in a generic bootstrap sample (for example, when $b = 1$). The results are summarized in tab.1. The different smoothers are sorted with respect to their number of parameters. The degree of the interpolating polynomial for the spline estimators and the magnitude of the smoothing parameter for the lowess estimators with one or two degrees of freedom are reported in brackets. The second column shows the number of parameters τ_k of each smoother whereas in the third we computed the outcome of the function $\nu(f_k)$. For each bootstrapped sample, our `bag.smooth.score` routine cycled through 15 alternative smoothers, and chose the right one after having sorted in ascending order the different fittings with respect to their level of parsimony (τ_k). The results underline that our approach privileges the linear estimator in the first case, by imposing a strong penalization to the less parsimonious smoothers, whereas in the second case the lowess estimator with 2 degrees of freedom and a bandwidth equal to 2/3 is preferred, despite the functions presenting higher degrees of smoothness provide better fits. These results are coherent with the hypothesis formulated when generating the two data structures.

Table 1. The smoother selection for a generic bootstrap sample

			sim1			sim2		
Smoother	τ_k	$\nu(f_k)$	$\phi[f_k(X_j, b)]$	$\rho_{1,k}$	$\rho_{100,k}$	$\phi[f_k(X_j, b)]$	$\rho_{1,k}$	$\rho_{100,k}$
intercept	1	0.933	0.000	0.000	0.000	0.00	0.000	0.00
linear	2	**0.867**	**0.991**	**0.101**	**1.00**	0.116	0.020	0.00
lowess(1, 0.66)	3	0.800	0.991	0.094	0.00	0.529	0.085	0.00
spline(3)	4	0.733	0.991	0.086	0.00	0.553	0.082	0.00
lowess(1, 0.5)	4	0.733	0.991	0.086	0.00	0.630	0.093	0.00
spline(4)	5	0.667	0.991	0.078	0.00	0.651	0.087	0.00
lowess(2, 0.75)	5	0.667	0.991	0.078	0.00	0.649	0.087	0.02
lowess (2, 0.66)	5	0.667	0.991	0.078	0.00	**0.691**	**0.093**	**0.62**
spline(5)	6	0.600	0.991	0.070	0.00	0.698	0.084	0.00
lowess (1, 0.33)	6	0.600	0.991	0.070	0.00	0.701	0.085	0.00
lowess (2, 0.5)	7	0.533	0.991	0.062	0.00	0.715	0.077	0.00
lowess (1, 0.25)	8	0.467	0.991	0.055	0.00	0.715	0.067	0.00
lowess (2, 0.33)	10	0.333	0.991	0.040	0.00	0.728	0.049	0.33
lowess (2, 0.25)	14	0.067	0.991	0.008	0.00	0.742	0.010	0.01

4 A Modified Backfitting Algorithm for Correct Model Identifiability

Once that we founded the most suitable smoother for each predictor, we are able to define a variable selection procedure for the definition of an ordering entrance of the predictors in the model. The different steps of the modified backfitting procedure, that we named *Smoothing Score* Algorithm, are summarized in table 2.
In the first step, the algorithm provides the sequence of ordered predictors $\{X_{(1)}, X_{(2)}, \ldots, X_{(s)}, \ldots, X_{(d)}\}$ to be possibly included in the model, whereas in the third step it produces the partial residuals in each sub-iteration (r), and evaluate the improvement in the fitting provided by the last predictor entered in the model according to the Anova-like F-test widely used in GAM modelling (see Hastie (1992, p.254) and Schimek (2000, p. 305)). Once that a predictor is included in the model, the partial residuals are computed for the sub-iteration $r+1$. The bagging procedure is applied on these partial residuals in order to select different smoothers for each candidate predictor. Repeating recursively step 3, the algorithm stops the variable selection procedure when the inclusion of

one among the $(d - s + 1)$ candidate predictors is not significant, as well as when all the d predictors are included in the model. Then, the standard backfitting procedure is applied until convergence for the model frame defined with the selected predictors.

Table 2. The Smoothing Score Algorithm

1. *Ordering.* Covariates are sorted decreasingly, according to the bagged smoother selections f_j^* defined in eq. 5.

2. *Initialize.* $i \leftarrow r \leftarrow f_s^{*(r)}(X_{(s)}) \leftarrow 0, \forall s = 1, \ldots, d$

3. *Update.* $1 \leftarrow r$
   For the s-th predictor $(s = 1, \ldots, d)$
   Fit the function $f_s^{*(r)}(X_{(s)})$ to the partial residuals
   $e^{(r-1)} \leftarrow [Y - f_s^{*(r-1)}(X_{(s)})]$.
   Test if the improvement in the goodness of fit of the overall model induced by the entrance of the s-th predictor is significant.
   Define a new ordering entrance by applying step 1 to the partial residuals:
   $$e^{(r)} \leftarrow [Y - f_s^{*(r)}(X_{(s)})]$$
   $r = r + 1$

4. *Verify.* Cycle step 3 and stop the variable selection procedure when any of the $(d - s + 1)$ predictors improves significantly the overall goodness of fit.

5. *Backfitting.* Apply standard backfitting to the residuals $[Y - \sum_{j=1}^{s} f_s^*(X_s)]$

5 A Comparison with Standard GAM

Our procedure for variable selection in GAM was applied to the Bupa Liver Disorders data set, contributed to the UCI Machine Learning Repository (Blake and Mertz, (1998)). The goal was to classify 345 individuals on the basis of two possible levels of alcohol consumption. In fact, the response variable is dichotomous and the initial error rate is 0.516, corresponding to the proportion of people presenting high alcohol consumption with respect to those classified as regular drinkers. The classification problem involves six predictors, one concerning the number of half-pint equivalents of alcoholic beverages drunk per day (drinks), the others consisting in blood tests which are thought to be sensitive to liver disorders that might arise from excessive alcohol consumption, namely the mean corpuscular volume (*mcv*), the alkaline phosphotase (*alkphos*), the alamine aminotransferase (*sgpt*), the aspartate aminotransferase (*sgot*) and the gamma-glutamyl transpeptidase (*gammagt*).

We use the `step.gam` approach of Hastie (1992) as a benchmark for our methodology and compare the relative results. In the `step.gam` approach, the model selection is made stepping through arbitrary models in a prespecified path, namely, once that the user expressly defines an *ordered regimen* of candidate smoothers for each predictor to be included in the model. The selected model is the one minimizing a modified Akaike Information Criterion (AIC).

For our experiment, in the variable selection step all the predictors were included in the GAM model for the two different approaches (values of the scoring functions are not showed here for sake of brevity). We compare the results of the final models, namely the estimation procedures following the AIC criterion and the *Smoothing Score* algorithm, in terms of the significance in the improvement of the goodness of fit provided by each predictor in the standard backfitting with respect to our modified backfitting. The results of the Anova-like F-Test for the two approaches are summarized in tab. 3. It is evident that our approach overperform standard GAM. In fact, the error rate provided by our approach based on bagging and smoothers scoring is lower than the one produced by standard GAM with respect to all the considered models, namely once that a further predictor is included in the model. At the end, the final error rate in our case is 0.201, whereas for the standard GAM is 0.370. Moreover, when adding a predictor in our model, the contribution it induces in the goodness of fit of the data expressed by the relative decrease in the Residual Sum of Squares is always significant (the p-values are always lower than 0.05). This is not the case of the standard GAM, where for the third predictor the p-value is 0.08.

Table 3. A comparison between the performance of the two different approaches.

Standard GAM					
Model	*n. of parameters*	*Residual Deviance*	*F*	*Prob.*	*Error Rate*
intercept	1	449.4	-	-	0.516
lowess(*mcv*)	4	458.7	2.84	0.02	0.504
alkphos	1	455.7	3.07	0.08	0.501
spline(*sgpt*)	4	440.2	4.15	0.00	0.484
Sgot	1	410.2	33.22	0.00	0.484
spline(*gammagt*)	4	355.7	13.96	0.00	0.391
spline(*drinks*)	4	337.8	4.73	0.00	**0.370**

GAM based on bagging and scoring					
Model	*n. of parameters*	*Residual Deviance*	*F*	*Prob.*	*Error Rate*
intercept	1	449.4	-	-	0.516
lowess(*drinks*)	4	439.6	5.24	0.00	0.406
lowess(*sgot*)	4	418.2	3.45	0.00	0.365
alkphos	1	402.6	2.53	0.02	0.319
lowess(*sgpt*)	4	377.1	6.16	0.00	0.264
spline(*gammagt*)	3	331.0	14.26	0.00	0.209
lowess(*mcv*)	4	316.0	3.87	0.00	**0.201**

6 Concluding Remarks

According to Hastie *et.* al (2001), GAMs assume a structured form for the unknown regression function, and by doing so they finesse the curse of dimensionality. Of course, they pay the possible price of misspecifying the model, and so in each case there is a tradeoff that has to be made. The proposed procedure for variable selection in GAMs try to handle with these problems.

Our approach uses the bagging procedure introduced in the framework of decision trees by Breiman (1996), who demonstrated how this procedure efficiently reduces the variance of regression pre-

dictors, while leaving the bias unchanged. In our case, bagging is used as a tool to improve the accuracy of the selection of the set of most powerful predictors (with the associated smoother); practically, we evaluate the sensitiveness of each variable and smoother selection with respect to small perturbations induced in the data. We have shown that our automatic procedure for variable and smoother selection in GAM improves the identification of the correct model and, consequently, the model fitting.
The theoretical motivation underlying this procedure is that with this approach the identification of the correct model is made in the first iteration of the backfitting algorithm. In this way, the identification and the estimation steps are not separated, opposed to the `step.gam` approach. The evaluation of the predictive accuracy of each candidate predictor based on the partial residuals of the models allows also to reduce the problems of multicollinearity and concurvity, that typically affect model fitting.
In the application, we achieved encouraging results in the case of a categorical response, but the *Smoothing Score* algorithm can be applied straightforwardly in the numerical response case. In general, the bagging and smoothing score approach can be fruitfully used when dealing with complex data structures and in the presence of a lot of predictors.
A wide discussion with some experiments about the effectiveness of the proposed scoring function for GAM-MM can be found in Conversano (2002).
Future works will be to extend this approach to the case of a multidimensional fit, namely when a single smoothing function is used to jointly model more than one predictor.

Acknowledgements. Author is grateful to anonymous referees and to Prof. Roberta Siciliano of the University of Naples Federico II for her insightful comments and remarks. The research was supported by MURST funds 1999 (awarded by R. Siciliano) and MURST funds 2001 (awarded by F. Mola).

References

BREIMAN, L. (1996). Bagging predictors, *Machine Learning*, **26**, 46-59.

BLAKE, C.L. and MERTZ, C.J. (1998). UCI repository of machine learning databases, *http://www.ics.uci.edu*.

CONVERSANO, C. (2002). Bagged Mixtures of Classifiers using Model Scoring Criteria, *Journal of Pattern Analysis and Applications*, **5**, 4, 351-362.

CONVERSANO, C. (2001). Generalized Additive Multi-Mixture Models: Scoring Models and Predictors, In: Klein, B. and Korsholm, L. (eds.), *New Trends in Statistical Modelling, Proceedings of the 16th International Workshop on Statistical Modelling*, University of Southern Denmark, 103-110.

CONVERSANO, C., SICILIANO, R. and MOLA, F. (2002). Generalized Additive Multi-Mixture Models for Data Mining, *Computational Statistics and Data Analysis*, **38**, 4, 487-500, special issue on Nonlinear Methods and Data Mining.

HASTIE, T., FRIEDMAN, J. and TIBSHIRANI, R. (2001). *Elements of Statistical Learning*, Springer, New York.

HASTIE, T. and TIBSHIRANI, R. (1990). *Generalized Additive Models*, Chapman & Hall, London.

HASTIE, T.(1992). Generalized Additive Models, In: Chambers, J. M. and Hastie, T. (eds.): *Statistical Models in S*, Wadsworth & Brooks/Cole Computer Science Series, Pacific Groove, California.

SCHIMEK, M.G. (2000). Additive and Generalized Additive Models, In: Schimek M.G. (Ed.), *Smoothing and Regression*, John Wiley & Sons, New York.

Bootstrap Variables Selection in Neural Network Regression Models

Francesco Giordano, Michele La Rocca, and Cira Perna

Department of Economics and Statistics, University of Salerno
Via Ponte Don Melillo, 84084 Fisciano (SA), Italy.
e-mail:[giordano,larocca,perna]@unisa.it.

Abstract. In this paper we consider the problem of variables selection in a non linear regression model with dependent errors. In this framework, we discuss the use of some measures for the variables relevance to the neural network model and we propose the use of the moving block bootstrap technique to estimate the variability of these measures. The performance of the procedure is evaluated by a small Monte Carlo experiment which shows how the proposed approach determines a correct ranking among relevant and irrelevant variables.

1 Introduction

Diagnostic and specification tests for model selection form the basis of a broadly accepted and coherent modelling approach, ensuring that the estimated model is a good approximation of the data generating process. In a regression framework, a key issue is variables selection, to avoid omission of relevant variables or inclusion of irrelevant ones. When using neural networks to estimate the regression function, testing whether the weights are significantly different from zero could be misleading since different network topologies can lead to similar approximation accuracy. Therefore, variables selection cannot be faced by focusing on single weights since, due the black-box nature of the network, it is difficult if not impossible to interpret them. Moreover, this approach can be inadequate in testing the overall significance of an explanatory variable and does not allow an understanding of its effect on the prediction of the model. From a statistical standpoint, it seems much more helpful to introduce a measure for the variable relevance and an estimate of its standard error in order to derive a formal statistical test.

In this paper we propose and discuss the use of moving block bootstrap to estimate the variability of some relevance measures for variable selection in neural network regression models with dependent errors. The paper is organized as follows. In the next section we briefly review artificial neural networks and in section 3 we focus on relevance measures and on the bootstrap procedure we propose. In section 4, we report the results of a Monte Carlo simulation experiment and in the last section some concluding remarks.

2 Neural Network Regression Models

Let $\{Y_t, t = 1, \ldots, T\}$ be a time series modelled as $Y_t = f(\mathbf{x}_t) + Z_t$, where f is a non linear continuous function, $\mathbf{x}_t = (x_{1t}, \ldots, x_{dt})$ is a vector of d non stochastic explanatory variables defined on $\aleph \subset \mathcal{R}^d$, and Z_t are (possibly) dependent random variables with zero mean. The function f can be approximated by a *single hidden layer feedforward network* of the form:

$$g(\mathbf{x}_t; \theta) = \sum_{k=1}^{m} c_k \phi \left(\sum_{j=1}^{d} a_{kj} x_{jt} + a_{k0} \right) + c_0$$

where $\theta = (c_0, c_1, \ldots, \mathbf{a}'_1, \ldots, \mathbf{a}'_m)$ with $\mathbf{a}'_k = (a_{k0}, a_{k1}, \ldots, a_{kd})$; c_k, $k = 1, \ldots, m$ is the weight of the link between the k-th neuron in the hidden layer and the output; a_{ki} is the weight of the connection between the j-th input neuron and the k-th neuron in the hidden layer; a_{k0} and c_0 are respectively the bias term of the hidden neurons and of the output; $\phi(\cdot)$ is the logistic function.

Neural network models can approximate any L^2 function and its derivatives to any level of accuracy (Cybenko, 1989; Hornik et al., 1990) and they become useful in high dimensional regression problems by looking for low dimensional decompositions (Barron, 1991). They are highly flexible and provide predictions with a relatively small bias. These good properties make neural networks natural candidates for modelling multivariate data but the choice of a proper topology for a given problem is still an open question. This requires choosing both an appropriate number of hidden units and a suitable set of explicative variables and, as a consequence, the connections thereof.

The most popular approaches used so far are regularization and pruning (Reed, 1993). In the first method, the network weights are chosen such that they minimize the network error function plus a penalty term for the network complexity. While, when using pruning methods, the aim is to identify those parameters that do not significantly contribute to the overall network performance. In both cases different topologies are chosen according to the single weights.

Although these methods may lead to satisfactory results, looking at the single weights can be misleading. This approach does not give any information on the most "significant" variables, which is useful in any model building strategy and different topologies can achieve the same approximation accuracy. Therefore, a proper choice of the network can be just based on complexity reasons and not on model plausibility.

All these techniques lack of a statistical perspective and are much more on the side of computational standpoint. Instead, it would be of some interest to look at the choice of the network topology in a statistical perspective, including this problem in the classical model selection approach (see Ripley, 1995, and the references therein). If focus is on diagnostic tests, Ardes and Korn (1999) propose a complete strategy based on a sequence of Lagrange multipliers tests on sets of weights. Although the statistical perspective is very strong, again, the role of the variables is not stressed.

To emphasize the role of the variables for the choice of the neural network topology a different strategy can be employed. Generally a desirable neural network contains as few hidden units as necessary for a good approximation of the true function, taking into account the trade-off between estimation bias and variability. As a consequence, once fixed the hidden layer size by using some asymptotic results (see Perna and Giordano, 2001) or some alternative *ad-hoc* procedures (Faraway and Chatfiled, 1998), attention can be focused on model identification in a classical sense:

the selection of the explanatory variables, in order to remove the "irrelevant" ones.

The approach we consider here is based on the introduction of some measures for the variable relevance which also, in an exploratory view, allows a ranking of the available explanatory variables. Following Refenes and Zapranis (1999) this involves (i) quantifying a "relevance" measure, (ii) estimating the sampling variability of this measure and (iii) testing the hypothesis of insignificance. In this paper we focus on the use of resampling techniques to estimate the standard error of some relevance measures. This choice is necessary in a neural network framework, where the use of asymptotic results if possible in principle, becomes soon very difficult and almost impractical in real problems. Of course, when dealing with dependent data, residual bootstrap is not consistent and some alternative bootstrap schemes should be considered.

3 Relevance Measures and the Moving Block Bootstrap for Variables Selection

In a linear regression model the relevance of a variable is measured by its coefficient which is also the magnitude of the partial derivative of the dependent variable with respect to the variable itself. So, in this set up, testing whether this coefficient is zero is equivalent to testing the hypothesis that the variable is relevant for the model. When dealing with nonlinear functions, the partial derivative is not a constant but it varies through the range of the independent variable x_j. Therefore, some kind of synthesis for the partial derivative over the range of x_j is needed.

A number of criteria, each measuring different aspects of relevance, can be used. The most popular one is the mean derivative, $M_j = Mean\,(D_{tj})$ where $D_{tj} = \partial\tilde{Y}_t/\partial x_{tj}$ is the change in $\tilde{Y}_t = g\,(\mathbf{x}_t;\theta)$ for a very small perturbation in the x_j. Alternatively, to avoid cancellations between negative and positive values, it can be used the mean absolute derivative $MA_j = Mean\,(|D_{tj}|)$ or the weighted quadratic mean $QM = \sqrt{Mean\left(D_{tj}^2\right)} \times \sigma_x/\sigma_z$, where σ_x

is the standard deviation of the independent variable and σ_z is the standard deviation of the residuals. In some applications, such as financial studies, it is also important to detect if $\tilde{Y}_t$ is sensitive to the independent variable x_j only in a small percentage of the input variable. So, the indexes: $MaxAD_j = \max_t (|D_{tj}|)$, $MaxD_j = \max_t (D_{tj})$, $MinAD_j = \min_t (|D_{tj}|)$, $MinD_j = \min_t (D_{tj})$ can also be considered.

Since all the relevance measures are functionals of the parameter vector, the sampling variability of their estimates could be computed analytically by a quite tedious process based on some asymptotic arguments. Alternatively, it can be used a feasible estimator based on stochastic simulations such as bootstrapping.

In the context of neural network regression models with *iid* errors, the residual bootstrap technique has been pursued in Tibshirani (1995) and Refenes and Zapranis (1999). When dealing with dependent data, this approach is not consistent and some modifications of the original procedure are needed in order to preserve the dependence structure of the original data in the bootstrap samples. For this aim, we propose to use a blockwise resampling scheme, the Moving Block Bootstrap (MBB), which gives consistent results under very general and minimal conditions and enjoys the properties of being robust against misspecified models (Kunsch, 1989; Politis e Romano, 1992 *inter alia*). This is particularly useful in neural networks where the specification of an adequate neural network topology can be difficult.

Moreover, even if the "best" topology is selected, neural networks are "atheoretical" and employed for the lack of knowledge about the form of the data generating process. Therefore, it is reasonable to consider a neural network as an approximation to an underlying model and analyze it as being misspecified in the sense of White (1994). Finally, estimation procedures for neural networks do not produce unique optimums and very different solutions are related to very similar values for the loss function and so the model cannot be considered globally identified.

The MBB algorithm can be adapted to possibly non stationary time series, in a neural network context, as follows.

Step 1. Compute the neural network estimates $g\left(\mathbf{x}_t;\hat{\theta}\right)$ for $t = 1,\ldots,T$ the residuals $\hat{Z}_t = Y_t - g\left(\mathbf{x}_t;\hat{\theta}\right)$ and the centered residuals $\bar{Z}_t = \hat{Z}_t - \sum_{t=1}^{T}\hat{Z}_t/T$.
Step 2. Fix $L < T$ and form blocks of length L of consecutive observations from the original data, the bootstrap sample is $Z^*_{(j-1)L+t} = \bar{Z}_{S_j+t}$, $1 \leq j \leq b$, $1 \leq t \leq L$ where $b = [T/L]$ denoting with $[x]$ the smallest integer greater or equal to x. Let $S_1, S_2, \ldots, S_b$ *iid* uniform on $\{0,1,\ldots,T-L\}$. If T is not a multiple of L, only $T+L-bl$ observations from the last block are used. The block length can be chosen as $L = \left[C \cdot T^{\lambda}\right]$ where $C > 0$ and $\lambda \in (0,1)$ with $[x]$ denoting the closest integer to x. When estimating a standard error $\lambda = 1/3$ and the constant C can be derived numerically (Hall et al., 1995) or by fixing it, as usual, at a value between 1/2 and 2.
Step 3. From the bootstrap replicate $\{Z_1^*,\ldots,Z_T^*\}$, generate the bootstrap replicate of the original series as $Y_t^* = g\left(\mathbf{x}_t;\hat{\theta}\right) + Z_t^*$ with $t = 1,\ldots,T$ and compute the bootstrap estimates of the neural network parameters, θ^*.
Step 4. Given one of the relevance measure introduced before, let's say $\Im = \Im(\theta)$, derive the bootstrap analogue, $\Im^* = \Im(\theta^*)$. Then , the variability of $\hat{\Im} = \Im\left(\hat{\theta}\right)$ is estimated by the bootstrap variance $var^*[\Im^*]$, where $var^*[\cdot]$ denotes the variance conditional on the observed data $(Y_t, \mathbf{x}_t)$ $t = 1,\ldots,T$. As usual, the bootstrap variance can be estimated by a Monte Carlo simulation.

4 Monte Carlo Results

A Monte Carlo experiment has been performed to illustrate the process of variables selection in the context of neural model identification. The data has been generated according to the Wahba's function as

$$Y_t = 4.26\left(e^{-x} - 4e^{-2x} + 3e^{-3x}\right) + Z_t$$

with $x \in [0, 2.5]$ and three different specifications for the error terms Z_t: a white noise; an $ARMA(1,1)$, specified as $Z_t =$

$-0.8Z_{t-1} - 0.5\epsilon_{t-1} + \epsilon_t$ with the innovations ϵ_t distributed as a Student-t with 6 degrees of freedom; an $EXPAR(2)$ model defined as

$$Z_t = \left(0.5 + 0.9 \cdot \exp\left(-Z_{t-1}^2\right)\right) \cdot Z_{t-1} - \\ - \left(0.8 - 1.8 \cdot \exp\left(-Z_{t-1}^2\right)\right) \cdot Z_{t-2} + \epsilon_t$$

with the innovations ϵ_t distributed as standard normal. The first model (M1) reproduces the usual regression model with *iid* errors; the second model (M2) is to consider a linear error process with non gaussian innovations with leptokutic distribution, a typical structure when considering some popular applications of neural networks such as those to financial data. Finally, the third model (M3) allows to consider both deterministic and stochastic nonlinear components.

The simulations are based on time series of length $T = 400$, 100 Monte Carlo runs and 50 bootstrap replicates. We limited the hidden layer size m in the set $\{3, 6\}$ since the Wahba's function can be approximated very well just with few neurons and values greater than 6 give serious overfitting problems. The block length L varies in the set $\{4, 6, 12\}$ in order to study its impact on the performance of the procedure. This parameter is critical in most applications since if the block of consecutive observations is not long enough, the dependence in the original series could not be reasonably preserved in the resampled ones.

We assume that the data generating process is unknown both in the functional form and in the relevant variables and we try to reproduce it by $\hat{Y} = \psi(x, w) + z$ where w is a white noise. The aim of the analysis is to show that the model identification procedure will correctly identify x as the most relevant variable. The sensitivity of the dependent variable to the two explanatory variables, x and w, is measured by the seven different relevance indexes defined in the previous section. They have been studentized as $t = \hat{\Im}/\sqrt{var^* [\Im^*]}$ so that the larger this value is the higher is the probability that the variable is relevant. The means over the 100 Monte Carlo runs are reported in the following tables.

The relevant variable is correctly identified as x by all the relevance measures with a clear cut between the two variables and

Table 1. Variable significance statistics: sensitivity of the dependent variable to the explanatory variables, x and w. White noise error process.

m	*L*	*Variable*	*M*	*MA*	*QM*	*MaxD*	*MaxAD*	*MinD*	*MinAD*
3	4	x	10.924	14.969	8.377	7.444	5.462	-1.757	0.901
		w	0.042	1.683	1.589	1.011	1.195	-1.043	0.389
	6	x	9.356	13.303	7.228	6.628	4.485	-1.690	0.935
		w	-0.151	1.536	1.320	0.857	0.950	-0.811	0.281
	12	x	7.837	11.822	6.532	6.823	4.163	-1.762	1.012
		w	0.150	1.974	1.758	1.204	1.390	-1.239	0.406
6	4	x	2.822	13.541	7.150	6.778	4.297	-4.077	0.717
		w	0.082	2.226	1.970	1.333	1.384	-1.275	0.548
	6	x	2.869	12.653	6.774	6.958	4.230	-3.786	0.620
		w	-0.122	2.352	2.106	1.432	1.454	-1.197	0.478
	12	x	2.743	12.157	6.090	6.346	3.803	-3.303	0.790
		w	0.054	2.239	1.942	1.223	1.306	-1.174	0.554

Table 2. Variable significance statistics: sensitivity of the dependent variable to the explanatory variables, x and w. ARMA error process with Student-t innovations.

m	*L*	*Variable*	*M*	*MA*	*QM*	*MaxD*	*MaxAD*	*MinD*	*MinAD*
3	4	x	15.306	25.474	10.820	12.890	9.385	-4.688	1.218
		w	0.019	1.722	1.549	1.087	1.280	-1.098	0.355
	6	x	12.543	25.225	9.867	14.128	9.571	-6.047	1.233
		w	-0.054	1.853	1.736	1.058	1.480	-1.331	0.353
	12	x	11.199	25.109	8.818	14.045	8.877	-5.736	1.011
		w	-0.001	1.887	1.707	1.379	2.610	-2.218	0.297
6	4	x	7.111	29.038	11.308	7.665	8.217	-7.542	0.817
		w	0.084	2.141	1.845	1.162	1.257	-1.137	0.531
	6	x	5.783	30.771	10.306	8.209	8.708	-8.080	1.005
		w	0.052	2.283	1.874	1.448	1.563	-1.225	0.548
	12	x	6.099	29.249	9.551	7.238	8.502	-8.226	0.960
		w	-0.131	2.100	1.720	1.362	1.304	-1.218	0.543

a unambiguous ranking comes out. Even in the white noise case, where no dependence among errors is present, the MBB works satisfactorily. In addition, for reasonable choices, the block length does not seem so critical.

The size of the hidden layer seems to have a greater impact on the value of the relevance measures. For $m = 6$ overfitting problem could arise and the discrimination between the relevant variable and the irrelevant one could be not clear as the case of $m = 3$.

Table 3. Variable significance statistics: sensitivity of the dependent variable to the explanatory variables, x and w. Expar error process.

m	L	*Variable*	M	MA	QM	$MaxD$	$MaxAD$	$MinD$	$MinAD$
3	4	x	12.064	17.051	9.497	8.928	6.145	-2.111	1.068
		w	-0.085	1.759	1.595	1.108	1.177	-0.994	0.318
	6	x	10.091	16.260	9.114	9.481	6.105	-2.951	0.905
		w	0.051	1.827	1.676	1.114	1.338	-1.100	0.333
	12	x	8.863	14.662	8.337	8.959	5.288	-2.543	0.855
		w	0.061	1.760	1.589	1.025	1.199	-1.023	0.269
6	4	x	3.856	17.499	9.767	7.108	5.789	-5.399	0.726
		w	0.140	2.046	1.747	1.132	1.181	-1.018	0.573
	6	x	3.530	16.563	8.965	7.949	5.399	-5.100	0.813
		w	-0.013	2.404	2.100	1.349	1.494	-1.331	0.579
	12	x	3.419	16.934	9.161	7.263	5.340	-5.030	0.662
		w	0.191	2.313	1.980	1.264	1.330	-1.168	0.607

In our simulations the mean values of the statistics considered for the two variables become closer but, in any case, the ranking is preserved.

Very similar results, not reported here, have been obtained by using more complex experimental designs by considering more than two explanatory, possibly correlated, variables. Again, the test is able to distinguish among relevant and irrelevant inputs.

To get a first impression on the distribution of the relevance measures, we focus our attention on the measure $M_j = Mean\,(D_{tj})$ since its simple structure suggests a possibly asymptotic Gaussian distribution. To assess departures from this model, in figure 1 we reported the Q-Q normal plots with envelope (Atkinson, 1985) of the sampling distribution of the relevance measure when $m = 3$, for the three models and for the different values of L. To produce the plots, we generated 19 samples of the same size from a normal distribution and we scale all samples to mean 0 and variance 1, in order to remove dependence on location and scale parameters. Then, each sample is sorted and for each order statistic, the maximum and the minimum values for the generated samples form the upper and the lower envelopes. Afterwards, the envelopes are plotted on the Q-Q plot of the scaled sampling distribution and

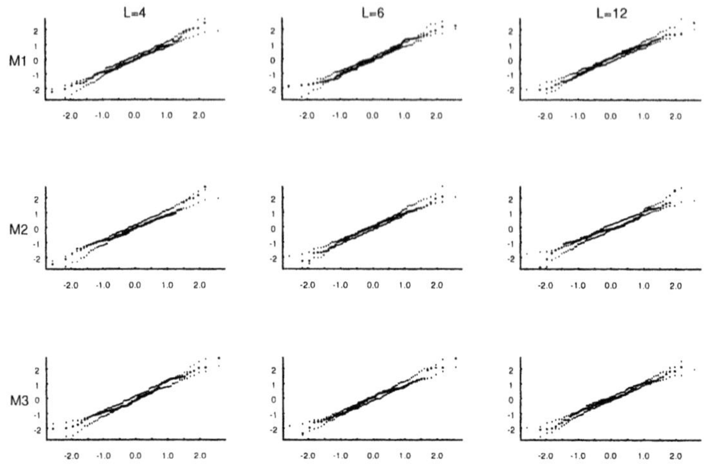

Fig. 1. Q-Q normal plots with envelope of the sampling distribution of the relevance measure $M_j = Mean\,(D_{tj})$ when $m = 3$, for the three models and for the different values of L.

form a guide to what constitutes serious deviations from the expected behavior under normality.

Looking at figure 1, it is quite clear that the Gaussian distribution is a reasonable approximation for the distribution of the test statistic and so it can be used as a guideline for significance testing. Therefore, as a rule of thumb, values of the statistic greater than 2 can be considered as significant at the usual level of 5% and the corresponding variable is assumed to be ”relevant”.

5 Concluding Remarks

Neural networks have been considered mainly ”‘black-boxes”’ and ”‘data-mining”’ tools without particular emphasis on the statistical issues related to model identification and interpretability. In this paper, we have focused on a methodology useful to identify explanatory variables in a regression framework with dependent errors. In particular, we have proposed the use of the moving block bootstrap to estimate the standard error of some relevance measures. The approach can be used as an exploratory tool to select a proper set of inputs for neural network models and it seems effective to produce a correct ranking among relevant and irrelevant variables.

The results of a small Monte Carlo study are quite encouraging but, to get a deeper insight, several different aspects should be further explored.

Firstly, if interest is in constructing significance test procedures based on the relevance measures, the sampling distribution of the test statistics under the null hypothesis (that the chosen variable is not relevant) is needed. For the mean derivative relevance measure case, a standard normal approximation seems to be a reasonable choice. In the general case, the derivation of the sampling distributions for model adequacy statistics is still an open issue.

Secondly, several alternative nonparametric resampling schemes could be employed to get a measure of sampling variability or to construct approximate testing procedures. The performance of these techniques should be evaluated and compared with the blockwise bootstrap proposed here.

ACKNOWLEDGEMENTS

We would like to thank two anonymous referees for their useful comments. The paper has been supported by MIUR and CNR funds.

References

ANDERS, U. and KORN, O. (1999): Model selection in neural networks. *Neural Networks*, *12*, 309–323.

ATKINSON, A.C. (1985): *Plots, Trasformations and Regressions.* Oxford University Press, Oxford.

BARRON, A. R. (1991): Complexity regularization with application to artificial neural networks. In: G. Roussas (Ed.): *Nonparametric Functional Estimation and Related Topics.* Kluwer Academic Publishers, pp. 561–576.

CYBENKO, G. (1989): Approximations by superpositions of a sigmoidal function. *Math. Control, Signals, Systems*, *2*, 303–314.

FARAWAY, J. and CHATFIELD, C. (1998): Time series forecasting with neural networks: a comparative study using the airline data. *Applied Statistics*, *47*, 231–250.

HALL, P., HOROWITZ, J.L. and JING, B. (1995): On blocking rule for the block bootstrap with dependent data. *Biometrika, 82*, 561–574.

HORNIK, K., STINCHCOMBE, M. and WHITE, H. (1990): Universal approximation of an unknown mapping and its derivatives using multilayer feedforward networks. *Neural Networks, 3*, 551–560.

KUNSCH, H.R. (1989): The jacknife and the bootstrap for general stationary observations. *The Annals of Statistics, 17*, 1217–1241.

PERNA, C. and GIORDANO, F. (2001): The hidden layer size in feed-forward neural networks: a statistical point of view. *Metron, LIX*, 217–227.

POLITIS, D.N. and ROMANO, J.P. (1992): A circular block-resampling procedure for stationary data. In: C. Page and R. LePage (Eds.): *Exploring the limits of the bootstrap.* Springer-Verlag, NY.

REED, R. (1993): Pruning algorithms - a survey. *IEEE transactions on Neural Networks, 4*, 740–747.

REFENES, A.P.N. and ZAPRANIS, A.D. (1999): Neural model identification, variable selection and model adequacy. *Journal of Forecasting, 18*, 299–332.

RIPLEY, B. D. (1995): Statistical ideas for selecting network architectures. In: B. Kappen and S. Gielen (eds.), *Neural Networks: artificial intelligence and industrial applications*, Springer, pp. 183–190.

TIBSHIRANI, R. (1995): *A comparison of some error estimates for neural network models.* Research Report, Department of Preventive and Biostatistics, University of Toronto.

WHITE, H. (1994): *Estimation, inference and specification analysis.* Cambridge University Press, Cambridge.

Robust Centre Location in Radial Basis Function Networks

Marilena Pillati and Daniela G. Calò

Dipartimento di Scienze Statistiche, Università di Bologna,
Via Belle Arti 41, I-40126 Bologna, Italy

Abstract. One of the most important and broadly accepted solutions for the centre selection problem in radial basis function networks is due to Moody and Darken (1989), who propose to cluster the input vectors via the k-means clustering method. In this paper, an alternative robust solution, based on the spatial median, is proposed and compared also to the k-medoids method (Kaufman and Rousseeuw, 1990). Some simulation studies in the context of classification problems show that, when outlying data are present, the solution we propose improves the network performance and allows to define a more parsimonious network.

1 Introduction

Artificial neural networks are a particularly rich class of non linear models developed by cognitive scientists, often used in regression, classification and time series prediction. This modelling approach is very useful for many practical problems when the underlying laws governing the systems from which the data are generated are unknown. In this context, radial basis function networks (RBF networks) provide an attractive alternative to other feedforward networks because of their faster learning capability.

RBF networks can be represented as single-hidden-layer feedforward neural networks, in which the activation functions of the hidden layer units are radially symmetric (*radial basis functions*), often Gaussians, each giving a significant response only in a local region of the p-dimensional input space (see Bishop, 1995 or Ripley, 1996 for more details). The transformation performed by the j-th output unit can be represented as a weighted sum of k basis functions, each computed by a hidden unit, as follows

$$g_j(\mathbf{x}) = \alpha_{j0} + \sum_{h=1}^{k} \alpha_{jh}\phi_h(||\mathbf{x} - \mu_h||) \qquad (1)$$

where $\mathbf{x} \in R^p$ is an input vector, ϕ_h is the h-th basis function with centre vector $\mu_h \in R^p$, α_{jh} is the weight of the connection between the h-th hidden unit and the j-th output one, and $||\cdot||$ denotes the Euclidean norm. The constant term α_{j0} is added to compensate for the difference between the average value over the data set of the basis function activations and the corresponding target values.

In the Gaussian case, the activation function of the h-th hidden unit is

$$\phi_h\left(||\mathbf{x} - \mu_h||\right) = \exp\left(-\frac{1}{2\sigma_h^2}||\mathbf{x} - \mu_h||^2\right) \tag{2}$$

It is defined by the centre μ_h and by the width or scale parameter $\sigma_h \in R^+$, which influences the degree of smoothness of the output function.

For fixed centres and fixed widths, the coefficients of the linear combination (1) can be estimated by least squares. This suggests to estimate the network parameters following a two steps procedure: in the first step the location and scale parameters of the basis functions are determined, in the second one the weights are estimated according to the least squares optimality criterion.

Of crucial concern is the first stage of the estimation process since the basis function centre locations strongly influence the network performance. Unless irrelevant variables are present (Pillati, 2001), the basis function centres should be learned by unsupervised procedures in order to reflect the data distribution in the input space.

One solution is to locate the centres at a collection of training elements, selected by random sampling or via subset selection or orthogonal least squares. As an alternative, the use of clustering techniques to find a set of centres which more accurately reflects the distribution of the data in the input space has been proposed (see Bishop, 1995, for a review).

The theoretical foundations of the clustering based estimation methods may be found in Poggio e Girosi (1990), who showed that a gradient descent approach used to update the RBF centres actually "... *makes the centres move towards the majority of the data, to find the position of the clusters"*. Therefore, basis functions with localized responses should be placed in regions of data concentration.

Moody and Darken (1989) proposed the k-means clustering algorithm (MacQueen, 1967) to optimally place RBF centres and a heuristic nearest neighbour method to determine the scale parameters. They estimated μ_h (h=1,...,k) through the cluster centroids, each basis function being representative of a cluster of training data. Unfortunately, this clustering method is sensitive to outlying data, thus yielding a set of locations which fails to reflect how the training set is scattered in the input space.

To overcome this problem, as an alternative to outlier deletion, we explore the possibility of robustly estimating the centres of the radial basis functions.

Among robust competitors of the mean vector in estimating location parameters in the multivariate context, we focus our attention on the so called *medoid* (Kaufman and Rousseeuw, 1990), and on the *spatial median* (Gower, 1974), both derived as "prototypes" in robust clustering algorithms. More precisely, we apply the *PAM* (*Partitioning Around Medoids*) algorithm to derive k medoids and propose a clustering algorithm which selects the cluster spatial medians as prototypes.

We evaluate the benefits of employing the proposed method by testing the performance of the spatial median based networks through some simulation experiments in the context of classification problems. The results, shown in the following, highlight that, when outlying data are present, the proposed solution improves the network performance and allows to define a more parsimonious classifier.

2 Multivariate Location Measures and Outlying Data

The k-means clustering method looks for a partition of the data set into k nonempty subsets and for k representative points that minimise the following objective function

$$J = \sum_{c=1}^{k} \sum_{i=1}^{n} u_{ci} ||\mathbf{x}_i - \mathbf{v}_c||^2 \tag{3}$$

where $\mathbf{v}_c \in R^p$ denotes the representative point or "prototype" of cluster c, and $u_{ci} \in \{0, 1\}$ indicates if $\mathbf{x}_i$ is a member of cluster c ($u_{ci} = 1$) or not. The problem is to minimise J with respect to the sets $U = \{u_{ci}\}$ and $V = \{\mathbf{v}_c\}$ under the following constraints

$$\sum_{c=1}^{k} u_{ci} = 1 \quad \forall i, i = 1, ..., n \tag{4}$$

$$0 < \sum_{i=1}^{n} u_{ci} < n \quad \forall c, c = 1, ..., k \tag{5}$$

The solution may be approximated by iteratively minimising the functional J in each set of grouped coordinates: first V is fixed, and J is minimised with respect to U; then, given these partial solutions, J is minimised with respect to V (Bobrowski and Bezdek, 1991). The procedure is repeated until there are no further changes in the grouping.

Denoted with $D(\mathbf{x}_i, \mathbf{v}_c)$ the "dissimilarity" between a data point $\mathbf{x}_i$ and a prototype $\mathbf{v}_c$ (in (3) $D(\mathbf{x}_i, \mathbf{v}_c)$ is the squared Euclidean distance), it may be shown that the solution to the first optimisation step consists in assigning any unit i to the cluster c^* if $D(\mathbf{x}_i, \mathbf{v}_{c^*}) = \min D(\mathbf{x}_i, \mathbf{v}_c)$, *i.e.* to the cluster whose prototype is nearest to $\mathbf{x}_i$. In the second step, the prototypes are optimally chosen by minimising the within clusters sum of dissimilarities: in (3) this yields the cluster centroids.

As all variance minimisation techniques, the k-means clustering algorithm performs poorly when outlying data are present, because of their effect on the value of the function J to be minimised.

More precisely, as it yields a set of k joint location least squares estimators, it tends to isolate "large outliers", which therefore will be taken as centres of some basis functions. Even if outlying data may be interesting in exploratory data analysis, since they may represent some meaningful isolated clusters, they may play a different role when clustering techniques are used in the basis function optimisation step. In fact, they may lead to locate some basis functions in low density regions of the input space, and not in the bulk of the data, where the system could benefit from them more. Moreover, even if outliers are not isolated by the clustering algorithm, they still will affect the centre estimates, moving some centroids away from the bulk of the corresponding clusters.

As a result, if outliers are present the k-means algorithm leads to a set of centres which fails to reflect how the training set is scattered in the input space. In order to cope with this problem, a possible approach is to detect and remove outliers before applying the clustering procedure (Lay and Hwang, 1999). As an alternative to outlier deletion, we explore the effectiveness of robust clustering methods to optimally locate RBF centres.

One of the most important contributions on robustness in cluster analysis is known as k-*medoids* method, implemented in the program PAM (*Partitioning Around Medoids*) described in Kaufman and Rousseeuw (1990).

It could represent an alternative solution to the problem of selecting a representative collection of training elements as basis function centres. In fact, the clustering algorithm is based on the search for k representative objects (called *medoids*) in the data set, which should represent various aspects of the structure of the data. As suggested from the authors, the algorithm is especially recommended if one is interested in the representative objects themselves.

This clustering solution is more robust with respect to outliers because it is based on the minimisation of an objective function J involving the Euclidean distance or the Manhattan distance instead of the squared Euclidean distance. Thus, the k-medoid approach searches for k representative objects $\mathbf{v}_c$ in the data set such that the function

$$J = \sum_{c=1}^{k} \sum_{i=1}^{n} u_{ci} D(\mathbf{x}_i, \mathbf{v}_c) \qquad (6)$$

is minimised, where D is the Euclidean distance or the Manhattan one. In (6) u_{ci} is set equal to 1 if $\mathbf{v}_c$ is the medoid closest to $\mathbf{x}_i$. The corresponding k clusters are found by assigning each remaining object to the nearest representative one. Thus, the average distance of the representative object to all the other elements of the same cluster is minimised. For this reason, such an optimal representative object is called *medoid* of its cluster. However, this leads to outlier isolation in most situations (Kaufman and Rousseeuw, 1990).

Even if we use the objective function (6) and allow the k representative objects to be replaced by general p-dimensional points (Cooper, 1963; Spath, 1980), the k joint location measures (k *medians*) we obtain are non robust too. In fact, although the median is a robust centralisation measure, the selection of k "joint" medians through this formulation is very unstable: it may be shown that the introduction of one sufficiently remote value implies the selection of such a value as one of the representative points (Cuesta-Albertos *et al.*, 1997).

Figure 1 shows the behaviour of these "joint" location measures, and the corresponding clustering process, on a artificial one-dimensional data set, with an obvious two cluster structure. While in (a) the clusters are perfectly detected and the two medians reflect the data structure, in (b) the clustering process is derailed by the presence of only one outlier, which is selected as one of the two representative points.

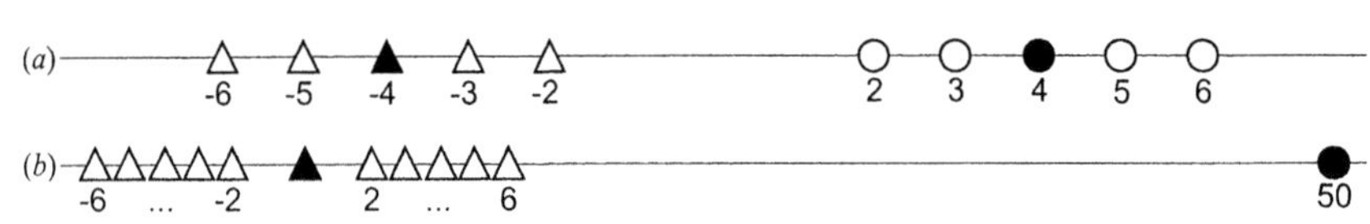

Fig. 1. Illustration of k-medians clustering ($k = 2$) in a one-dimensional data set (Garcia-Escudero et al. 1997). Different simbols (square and dot) denote the membership to different clusters, and the representative points are black.

3 The k-Spatial Median Estimates of the RBF Location Parameters

Taking into account that medians are robust competitors of the mean vector in estimating location parameters in the one-sample problem, we propose an iterative relocation algorithm, which is based on a particular notion of multidimensional median.

As an alternative to the exhaustive search involved in the minimisation of (6), we suggest to approximate the solution by iteratively minimising the objective function in each set of grouped coordinates, as in the k-means algorithm. As it will be explained in the following, first U is fixed, and J is minimised with respect to $\{\mathbf{v}_c\}$; then, given these partial solutions, J is minimised with respect to $\{u_{ci}\}$ by assigning any unit i to the cluster whose prototype is nearest to $\mathbf{x}_i$.

Given U, the prototypes are optimally chosen by minimising the within cluster sum of dissimilarities, but a different choice for $D(\mathbf{x}_i, \mathbf{v}_c)$ leads to a different set of representative objects. If $D(\mathbf{x}_i, \mathbf{v}_c)$ is the Manhattan distance the solution is given by the cluster median vectors; if $D(\mathbf{x}_i, \mathbf{v}_c)$ is the Euclidean distance the second step solution leads to the cluster *spatial medians*.

In fact, given a set of n points in R^p, $\mathbf{x}_1, \mathbf{x}_2, ..., \mathbf{x}_n$, its spatial median is the vector $\widehat{\theta}$ that minimises the sum of the Euclidean distances to the points, in direct analogy to the univariate sample median, that is

$$\widehat{\theta} = \arg\min \sum_{i=1}^{n} ||\mathbf{x}_i - \theta|| \tag{7}$$

We focus our attention on it for several reasons. It is invariant to rotations of the axes and thus preferable to the non invariant vector of the marginal medians. A geometrical interpretation shows that the spatial median is naturally applicable to spherically symmetric models (Small, 1990). Furthermore, following Vardi and Zhang (1999), the spatial median of a p-dimensional data cloud can be computed through a fast algorithm, which is extremely simple to program and is guaranteed to converge monotonically to the solution from any starting point in R^p.

As any iterative relocation algorithm, the one we propose requires an initial starting point. Starting from a random partition of the training set, and not from a random sample of k objects, the robustness of the spatial median guarantees that the extreme values are never selected as prototypes during the iterations of the procedure.

Figure 2 shows the behaviour of the iterative algorithm in the same situation reported in Figure 1(b). As opposite to the k-means clustering, the proposed algorithm does not reach the global solution of the minimisation problem, which corresponds to forcing one median to catch the outlier. It stops in a local minimum, yielding a set of cluster prototypes which represents a robust solution to the RBF centre location problem. As it can be seen in Figure 2, the iterative algorithm leads to the same representative points obtained by the k-means algorithm or by *PAM* when no outlier is present (Figure 1(a)).

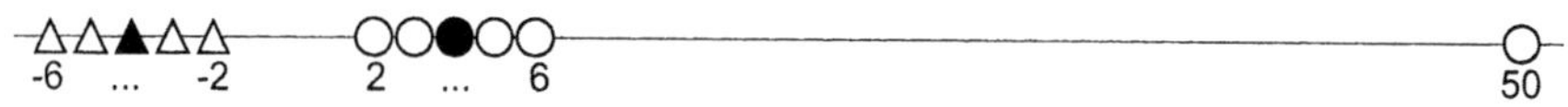

Fig. 2. Output of the proposed clustering procedure in the artificial data set of Figure 1(b).

4 Simulation Study and Concluding Remarks

The robustness of the proposed method against outlying observations has been tested on several data sets in the context of classification problems. Its performances have been compared to those of two RBF networks, which differ in the centre estimation method: the k-means based classifier and the k-medoids based one.

The estimation of the scale parameters followed a nearest neighbour based method, as suggested in Moody and Darken (1989). The procedures have been implemented in S-PLUS, resorting to the algorithm proposed by Vardi and Zhang (1999) in the computation of the spatial median and to the S-PLUS function *PAM* in the selection of the medoids.

The first simulation experiment deals with a 2-class problem in 6 dimensions. Class 1 is drawn from a uniform mixture of a multivariate unit normal with vector mean $(a, ..., a)$ and a multivariate unit normal with vector mean $(-a, ..., -a)$, with $a = 2/\sqrt{6}$. Class 2 is generated from a unit normal with vector mean $(a, -a, ..., a, -a)$.

A contaminated setting was simulated by drawing 10% of the observations in Class 2 from a distribution which differs from the original one only in the standard deviation (10 times the previous one).

Both in the non contaminated situation and in the contaminated one, 50 training sets of 300 observations each were generated. The results of these two simulation experiments are illustrated in Table 1 and Table 2, respectively. They show, for each classifier, the mean and the standard deviation of the 50 misclassification error rates evaluated on a test set of 3000 observations.

Table 1. Mean error rates and standard errors (in brackets) of RBF networks with different centre estimation methods in the 2-class problem (no outlying observation).

k	*k-means RBF*	*k-medoids RBF*	*k-spatial medians RBF*
2	0.3305 (0.0425)	0.3050 (0.0612)	0.3270 (0.0242)
3	0.1286 (0.0034)	0.1582 (0.0250)	0.1269 (0.0027)
4	0.1260 (0.0036)	0.1317 (0.0082)	0.1237 (0.0043)
5	0.1244 (0.0045)	0.1281 (0.0088)	0.1232 (0.0048)
6	0.1228 (0.0051)	0.1248 (0.0061)	0.1209 (0.0047)
7	0.1215 (0.0051)	0.1220 (0.0065)	0.1200 (0.0050)
8	0.1188 (0.0046)	0.1208 (0.0065)	0.1163 (0.0053)
9	0.1173 (0.0047)	0.1196 (0.0054)	0.1163 (0.0048)
10	0.1172 (0.0050)	0.1172 (0.0055)	0.1150 (0.0040)

The results show that in absence of outlying observations all the classifiers yield similar error rates for various k values, even if the choice of a subset of training elements through *PAM* seems to slightly worsen the network performances, unless k=2 where the three classifiers perform poorly. Due to the three-modal population structure, the use of more than three basis functions does not reduce sensibly the test set error rates for all the classifiers.

As expected, the k-means based RBF network proves to be derailed by outlying observations, while the estimate solution we propose seems to be affected by outliers to a lesser extent. Although the performances of both classifiers get better as the number k of hidden units increases, the spatial median based network achieves good results with a simpler architecture. As in the non contaminated case, as few as three basis functions are enough to get performances quite similar to those obtained with more hidden units.

Table 2. Mean error rates and standard errors (in brackets) of RBF networks with different centre estimation methods in the 2-class problem (10% of outlying observations in Class 2).

k	*k-means RBF*	*k-medoids RBF*	*k-spatial medians RBF*
2	0.4054 (0.0524)	0.3467 (0.0769)	0.3821 (0.0535)
3	0.2915 (0.1244)	0.1685 (0.0426)	0.1382 (0.0050)
4	0.2019 (0.0834)	0.1538 (0.0205)	0.1322 (0.0060)
5	0.1550 (0.0395)	0.1460 (0.0176)	0.1280 (0.0063)
6	0.1367 (0.0187)	0.1426 (0.0153)	0.1275 (0.0058)
7	0.1313 (0.0066)	0.1391 (0.0132)	0.1274 (0.0061)
8	0.1272 (0.0058)	0.1351 (0.0106)	0.1266 (0.0065)
9	0.1257 (0.0056)	0.1330 (0.0107)	0.1246 (0.0062)
10	0.1238 (0.0057)	0.1303 (0.0106)	0.1248 (0.0064)

Table 3 and Table 4 report the results of another simulation study, dealing with the "waveform data" (Breiman *et al.*, 1984). It is a 3-class problem in 21 dimensions. The classes have equal probability. The test set consists of 3000 observations. Fifty training sets of 600 observations each were generated. The misclassification error rate of each classifier has been estimated by averaging the 50 error rates on the test set.

In the contaminated setting (Table 4), 10% of the observations in each class has been generated from a distribution which differs from the original one only in the standard deviation (15 times as much as it was at the beginning).

Table 3. Mean error rates and standard errors (in brackets) of RBF networks with different centre estimation methods in the 3-class (no outlying observation).

k	k-*means RBF*	k-*medoids RBF*	k-*spatial medians RBF*
2	0.3712 (0.0088)	0.3712 (0.0217)	0.3659 (0.0095)
3	0.1637 (0.0107)	0.1828 (0.0134)	0.1649 (0.0113)
4	0.1597 (0.0137)	0.1862 (0.0119)	0.1601 (0.0099)
5	0.1470 (0.0087)	0.1798 (0.0110)	0.1490 (0.0120)
7	0.1394 (0.0045)	0.1690 (0.0080)	0.1400 (0.0050)
9	0.1397 (0.0040)	0.1634 (0.0065)	0.1393 (0.0038)
11	0.1400 (0.0041)	0.1597 (0.0068)	0.1403 (0.0041)
13	0.1406 (0.0047)	0.1579 (0.0059)	0.1401 (0.0048)
15	0.1412 (0.0044)	0.1554 (0.0062)	0.1407 (0.0052)

Table 4. Mean error rates and standard errors (in brackets) of RBF networks with different centre estimation methods in the 3-class problem (10% of outlying data).

k	k-*means RBF*	k-*medoids RBF*	k-*spatial medians RBF*
2	0.3767 (0.0108)	0.3841 (0.0144)	0.3726 (0.0080)
3	0.2542 (0.0782)	0.1928 (0.0218)	0.1814 (0.0166)
4	0.2190 (0.0744)	0.1861 (0.0197)	0.1779 (0.0206)
5	0.1836 (0.0466)	0.1818 (0.0154)	0.1570 (0.0180)
7	0.1703 (0.0271)	0.1796 (0.0145)	0.1431 (0.0055)
9	0.1663 (0.0179)	0.1796 (0.0138)	0.1412 (0.0050)
11	0.1678 (0.0168)	0.1787 (0.0130)	0.1402 (0.0047)
13	0.1626 (0.0108)	0.1779 (0.0126)	0.1400 (0.0039)
15	0.1584 (0.0142)	0.1785 (0.0115)	0.1406 (0.0042)

This simulation experiment confirms that in the non contaminated setting the k-means based classifier and the spatial median based one have similar performances, but the latter outperforms the former in presence of extreme observations. Due to the fact that the RBF centres must be selected from the training set, the k-medoid based classifier seems to be less flexible than the others. However, if few hidden units are considered, the presence of outliers makes it preferable to the k-means based classifier.

The simulation results seem therefore to confirm that, when outlying data are present, the solution we propose improves the network performance, even with regard to its stability, and allows to define a more parsimonious network.

References

BISHOP, M.C. (1995): *Neural Networks for Pattern Recognition.* Clarendon Press, Oxford.

BOBROWSKI, L. and BEZDEK, J.C. (1996): c-Means Clustering with the L_1 and L_∞ Norms. *IEEE Transactions on Systems, Man and Cybernetics, 21*, 545–554.

BREIMAN, L., FRIEDMAN, J., OLSHEN, R., and STONE, C. (1984): *Classification and Regression Trees.* Wadsworth and Broooks/Cole, Monterey.

COOPER, L. (1963): Location-Allocation Problems. *Operational Research, 11*, 331–343.

CUESTA-ALBERTOS, J.A., GORDALIZA, A., and MATRAN, C. (1997): Trimmed k-Means: An Attempt to Robustify Quantizers. *Annals of Statistics, 25*, 553–576.

GARCIA-ESCUDERO, L.A.,GORDALIZA, A., and MATRAN, C. (1997): k-Medians and Trimmed k-Medians. *Student, 2*, 139–148.

GOWER, J.C. (1974): The Mediancentre. *Applied Statistics, 23*, 466–470.

KAUFMAN, L. and ROUSSEEUW, P.L. (1990): *Finding Groups in Data.* Wiley, New York.

LAY, S.R. and HWANG, J.N. (1993): Robust Construction of Radial Basis Function Networks for Classification. *IEEE International Conference on Neural Networks*, 1859–1864.

MACQUEEN, J. (1967): Some Methods for Classification and Analysis of Multivariate Observations. In: L.M. LeCam and J. Neyman (Eds.), *Proceedings of the 5th Berkeley Symposium on Mathematical Statististics and Probability, 1*, 281–297.

MOODY, J. and DARKEN, C.J. (1989): Fast Learning in Network of Locally-Tuned Processing Units. *Neural Computation, 1*, 281–294.

PILLATI, M. (2001): Le Reti Neurali in Problemi di Classificazione: una Soluzione Basata sui Metodi di Segmentazione Binaria. *Statistica, 3*, 407–421.

POGGIO, T. and GIROSI, F. (1990): Networks for Approximation and Learning. *Proceeding of the IEEE, 78*, 1481–1497.

RIPLEY, B.D. (1996): *Pattern Recognition and Neural Networks.* Cambridge University Press, Cambridge.

SMALL, G. (1990): A Survey of Multidimensional Medians. *International Statistical Review, 58*, 263–277.

SPATH, H. (1980): *Cluster Analysis Algorithms.* Ellis Horwood Limited, Chichester.

VARDI, Y. and ZHANG, C. (1999): The Multivariate L_1-Median and Associated Data Depth. *Proceedings of the National Academy of Sciences of the United States of America. 97*, 1423–1426.

The Genetic Algorithm Estimates for the Parameters of Order p Normal Distributions

Salvatore Vitrano and Roberto Baragona

Dipartimento di Sociologia e Comunicazione, Università di Roma "La Sapienza," Via Salaria 113, I-00198 Roma, Italy

Abstract. The exponential power function distributions are a natural generalization of the normal distribution. A slight arrangement of their parameters originates the normal distributions of order p. The two formulations are fairly equivalent, and we shall use the second one because many properties, related to scale and shape, for instance, may be written in a simpler way. The parameter estimation of such distributions, however, is rather difficult in practice as non-linear equation systems have to be solved. In addition, the estimates that may be defined are often either biased or have large variance. In this paper we introduce a genetic algorithm for estimating the location and scale parameters, in the presence of small sample size, when the shape parameter p is assumed known. Then, if the sample size is moderate, a procedure based on genetic algorithms for estimating simultaneously the location, scale and shape parameters will be developed. A simulation experiment is carried out to illustrate the advantages and the merits of the proposed procedure compared with some well established methods.

1 Introduction

Subbotin (1923), generalizing the second Schiaparelli axiom, introduced the class of distributions named exponential power functions (EPF) as a generalized random error structure. Several formulations of the EPF were extensively analyzed in the literature (Diananda, 1949; Box, 1953; Turner, 1960; Box and Tiao, 1973), but here we refer to the model known as normal distributions of order p (Vianelli, 1963; Lunetta, 1963)

$$f(x; \mu, \sigma_p, p) = \left\{2p^{1/p}\sigma_p \Gamma(1 + 1/p)\right\}^{-1} \exp\left\{-\left(p\sigma_p^p\right)^{-1} |x - \mu|^p\right\}, \tag{1}$$

where $-\infty < x < +\infty$, $-\infty < \mu < +\infty$, $0 < \sigma_p < \infty$, $0 < p < \infty$. The density (1) is completely specified by the location μ, the scale σ_p and the shape parameter p. A family of unimodal

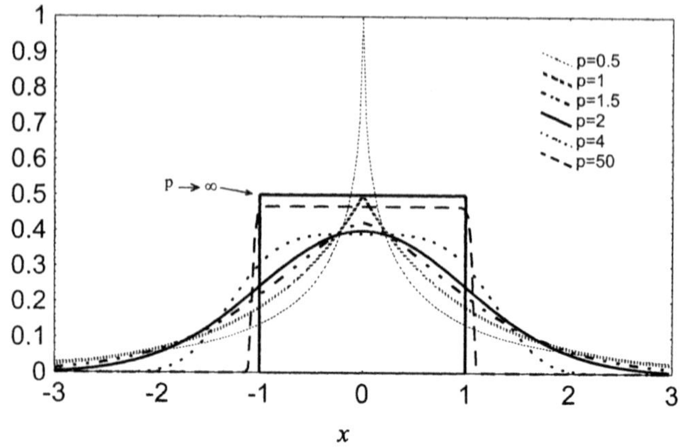

Fig. 1. Example of particular cases for normal distributions of order p. The location parameter is assumed $\mu = 0$, and the scale $\sigma_p = 1$.

symmetric curves from the density function (1) arise for different values of p, with shape varying from leptokurtic ($0 < p < 2$) to platykurtic ($p > 2$) distributions. The Laplace ($p = 1$), the Normal ($p = 2$), and the Uniform ($p \to \infty$) distributions may be obtained as special cases. In the Figure 1, some curves are plotted where different values for p were assumed, while μ was set to zero and σ_p was set to unity.

Let $\mathbf{x}=(x_1, x_2, \ldots, x_n)$ be an i.i.d. random sample drawn from this probability density function. The maximum-likelihood (ML) estimator of the parameter vector $\theta=(\mu, \sigma_p, p)$ is any value θ which maximizes the likelihood function

$$L(\theta, \mathbf{x}) = \left\{2p^{1/p}\sigma_p \Gamma(1+1/p)\right\}^{-n} \exp\left\{-\left(p\sigma_p^p\right)^{-1} \sum_{i=1}^{n} |x_i - \mu|^p\right\}. \tag{2}$$

Properties and performances of the parameter estimation methods for the distributions in this class were extensively studied by Mineo (1989), Mineo (1994), Agro' (1995), Burgio and Nikitin (1998), Chiodi (2000). This paper proposes a genetic algorithm (GA) for estimating θ based on the $L(\theta, \mathbf{x})$ function. Such approach is believed effective to avoid several difficulties that occur when the common search numerical algorithms are used. Alternative estimation procedures, that we do not consider here, were proposed which combine the ML and moment estimate methods based on kurtosis index. These methods were found able to give well-known optimal results also for small samples (Mineo, 1994). Two scale parameter estimators (Chiodi, 1988; La Tona, 1994) were introduced to reduce the bias of the ML scale estimator when p was supposed known. In fact, the usual correction of the degrees of freedom for variance computation does not work for $p \neq 2$, as the location parameter is replaced by its ML estimate.

Holland (1975) studied the evolution of a population of individuals in a given environment by means of a class of analytical models called GA. According to this approach, each individual in the population, which has not to be given a strictly statistical meaning, needs to be characterized by a chromosome, i.e. a string of characters. Each character in the string possesses a specific meaning that may be decoded from both its value, called allele, and position, called locus. The population evolves, by different genetic operators, through several generations, towards the best adaptation to the environment. A basic assumption is that a function may be defined that maps the genetic pool (the set of all admissible strings) into the positive real axis. This function is known as *fitness function* (ff). The ff is to be computed for each individual, and it has to increase as soon as the adaptation to the environment increases. The special applications of the GA for function optimization assume the objective function as the ff, and the string owned by the best fit individual as the solution to the optimization problem. The effectiveness of the GA for optimization may be enforced by using the so-called elitist strategy, where the best individual fitness function value is always recorded, and possibly updated in each generation. This approach is similar to the simulated annealing (SA) application to function optimization (Brooks and Morgan, 1995). Using GA for statistical inference was investigated by Chatterjee and Laudatto (1997).

The outline of the paper is as follows. The next section is devoted to the illustration of the GA we implemented for assessing estimates and standard errors. Comparison with alternative approaches by means of a simulation experiment is provided in

Section 3. The estimation of the location and scale parameters in case of small samples, for given order p, and the estimation of all three parameters, for rather large sample sizes, is considered. Conclusions are drawn in the last section.

2 Genetic Algorithms Implementation

The main features of our GA design are the tournament selection for reproduction, and three operators, crossover, inversion, and mutation. Unlike many GA applications, inversion proved to constitute a valid support to increase the speed of convergence to the solution. Encoding the potential solutions, that are real numbers in a given interval (a, b), say, was done by using the following device. Let us choose a positive integer ℓ as the binary string length. Then, 2^ℓ different numbers x may be coded according to

$$x = a + c(b - a)/(2^\ell - 1), \tag{3}$$

where c represents the real value corresponding to the given binary string. As c may vary from 0 (the null string) through $2^\ell - 1$ (the all-one string), then (3) encodes a finite subset of real numbers in (a, b), at equispaced intervals of length $(b - a)/(2^\ell - 1)$ each. The code (3) was used for each of the three parameters to be estimated. The integer ℓ was taken fixed, whilst three intervals (a, b) were selected according to the parameter constraints (e.g. p has to be positive), some preliminary plots of artificial data and suggestions from previous studies (see, for instance, Agro' (1995; 1999)).

Reproduction. Let the so-called pressure selection probability p_s be chosen which determines the rate at which individuals with high fitness spread into the population. A set of s potential solutions are generated at random as binary strings of length ℓ. We chose $\ell = 32$ in our application. The s individuals in the population are paired at random, then each pair is examined. A copy of the string of the individual for which a fitness function greater than the other is computed replaces the string of the less fit individual with probability p_s. So, there is a chance of $1 - p_s$ that the two individuals remain unchanged. Note that, if s is odd, an individual is left alone and remains unchanged.

Crossover. Though many crossover operators were available, for problems of this kind the single-point crossover seemed the best choice. The s individuals, after the reproduction step, are again paired at random, excepted one at most if s is odd. This latter remains unchanged as before. A positive integer into the range $(1, \ell - 1)$, j say, is selected at random. Then the characters from the $(j + 1)$-th are exchanged between the two individuals. The crossover operator applies to a pair with probability p_c.

Inversion. Each of the s individuals, after crossover took place, is considered in turn, and its string is processed by the inversion operator with probability p_i. Now, within the range $(1, \ell)$, two cut points are selected at random, and the characters in between are to be taken in reverse order.

Mutation. All characters in the whole genetic pool are allowed to change their values with probability p_m. The choice of p_m was proven to influence the performance of the GA considerably. A high p_m may serve to maintain the diversity between the individuals into the population, but this is likely to produce premature convergence as well.

For the present problem, most figures recommended by Chatterjee and Laudatto (1997) were assumed, i.e. $p_s = 0.75$, $p_c = 0.7$, $p_i = 0.65$, and 50 as the number of generations. We considered instead the population size rather small, $s = 50$, as no substantial improvement was observed by taking larger values. Figures were suggested for p_m in the range 0.001 through 0.1, but a right balance for the present problem seemed $p_m = 0.01$. We also assumed the $L(\theta, \mathbf{x})$ in (2) as the ff. The vector estimate θ was decoded from the chromosome of the best individual found by the GA.

3 Simulation and Results

Two different Monte Carlo simulation studies were designed, where the shape parameter p was distinctively supposed known and unknown. Through an algorithm of pseudo-random numbers (Chiodi,

1986), 100 samples simulated of size $n = 10, 20, 50$, for p supposed known, and of sample size $n = 50, 100$, with p unknown, were generated from a standardized normal distribution of order p, with $p = 1, 1.25, 1.5, 1.75, 2, 2.25, 2.75$ and 3.25. Each standardized value was afterwards consequently transformed, by considering a normal distribution of order p with θ=$(3, 2, p)$. We have considered larger sample sizes in the p unknown case because of the difficulty to reach the maximum of the function (2), for sample size $n \leq 100$ and for some p estimated values (Agro', 1995). For p known, the vector θ to be estimated will be composed by μ and σ_p. We also underline that even if the hypothesized situation of "p known" is unreliable one, we took it into account because many results, concerning, for instance, approximated sampling test distributions (Lunetta, 1966; Chiodi, 1994 and 2000), are still mainly based on this assumption.

As far as the first simulation experiment is concerned, four algorithms were compared. The first one is the ML procedure proposed by Chiodi (1988). The second one is essentially the jackknife estimator described in La Tona (1994). Then, the GA was used. The fourth algorithm is a SA procedure which is aimed at providing several starting points for a quasi-Newton algorithm. It was designed as a numerical optimization tool for maximizing the likelihood function. The results are displayed in Table 1. The sample means and the standard errors of the estimated vector θ were computed for each p value and sample size by using the procedures listed above. The least absolute bias, averaged over all the (p, n) pairs, for the location parameter μ, is obtained by the ML algorithm. The GA minimizes instead the average absolute bias for the scale parameter σ_p. It seems also interesting to distinguish the results according to the sample size. For $n = 10$, the least absolute bias, averaged over all values of the shape parameter p, may be computed from the GA estimates of both μ and σ_p. Nonetheless, the least standard errors, on the average, may be obtained, for μ, by the jackknife estimator, and, for σ_p, by the ML one. As far as the sample size $n = 20$ is concerned, the least average absolute bias and standard errors may be obtained by using the jackknife estimator, for μ, and by the GA, for σ_p. In case of sample size $n = 50$, for both location and shape parameters, the average absolute bias is minimized by the ML estimator, and the standard error of the estimates by the GA. Another kind of comparison may turn of interest by distinguishing the values of the shape parameter p in two sets, the first one including the figures from 1 through 2, the second one those from 2.25 through 3.25. In the latter case, the GA was able to give, by considering the average absolute bias, the best results for both parameters. In the former one, the ML algorithm provided best results for μ, and the GA for σ_p.

The results of the simulations, for p unknown, are given in Table 2, where the sample means and the standard errors of the estimated vector θ were computed for each p value and sample size, by using the ML and the GA approaches. The number K of samples, for which the ML procedure has not reached any optimum point of (2), is also reported in the last column. For all sample sizes and values of p, the sample mean of μ and σ_p are very close to their true values defined in θ for each approach. More precisely, the results obtained by the GA approach are more likely than the ML approach when $p > 2$, except some rare cases. The same conclusions do not work always for $p \leq 2$. About the shape parameter p, its true values were over estimated by its sample mean for both approaches, even if the results by the GA approach were always less biased than the ML approach, for p greater than 2. The results for $p \leq 2$, however, have to be interpreted with caution, because of the samples discarded by the ML procedure. No substantial differences about the standard errors of the estimates were found. Furthermore, we tried a SA procedure on the same problem (p unknown), and similar results were obtained.

4 Conclusions

Some procedures for the estimation of the parameters of an order p normal distribution are compared by means of two distinct Monte Carlo simulations. The numerical inspection of the sample

Table 1. Order-p normal distribution parameter estimates, p known

		Max. likeli.		Jackknife		Genetic alg.		Sim. anneal.	
p	n	μ	σ_p	μ	σ_p	μ	σ_p	μ	σ_p
1	10	2.9917	1.8810	2.9883	2.1404	3.0100	1.9800	2.9641	2.3512
		(.6917)	(.5415)	(.6011)	(.6393)	(.7275)	(.5700)	(.7062)	(.6769)
	20	2.9819	1.9523	2.9819	2.0684	2.9815	2.0116	2.9562	2.1792
		(.4879)	(.3587)	(.4879)	(.3777)	(.5410)	(.3492)	(.5342)	(.3783)
	50	3.0265	1.9948	3.0265	2.0426	3.0047	2.0157	3.0027	2.0787
		(.2841)	(.2475)	(.2841)	(.2665)	(.2937)	(.3036)	(.3021)	(.3131)
1.25	10	3.0241	2.0346	3.1013	2.0271	2.9859	1.9673	2.9860	2.1479
		(.7358)	(.5960)	(.6526)	(.5932)	(.7049)	(.5429)	(.7049)	(.5927)
	20	3.0292	2.0314	3.0283	2.0361	3.0434	1.9615	3.0430	2.0442
		(.5014)	(.4402)	(.4983)	(.4421)	(.5988)	(.4391)	(.5989)	(.4576)
	50	3.0088	2.0031	3.0100	2.0048	3.0350	2.0079	3.0356	2.0402
		(.3397)	(.2641)	(.3391)	(.2631)	(.2533)	(.2473)	(.2535)	(.2513)
1.5	10	3.0684	1.9882	3.0709	1.9907	3.0824	1.9685	3.0823	2.0559
		(.5888)	(.5845)	(.5906)	(.5860)	(.6297)	(.5474)	(.6296)	(.5717)
	20	3.0068	2.0271	3.0071	2.0277	3.0408	2.0082	3.0407	2.0498
		(.4896)	(.3651)	(.4906)	(.3675)	(.5124)	(.3879)	(.5124)	(.3960)
	50	2.9763	1.9748	2.9764	1.9750	3.0307	1.9605	3.0306	1.9761
		(.3028)	(.2428)	(.3024)	(.2423)	(.3195)	(.2501)	(.3194)	(.2513)
1.75	10	3.0205	1.9401	2.9829	1.9340	2.9185	1.9898	2.9185	2.0240
		(.6755)	(.4805)	(.5856)	(.4726)	(.6552)	(.4784)	(.6552)	(.4866)
	20	3.0358	1.9979	3.0360	1.9987	2.9739	1.9959	2.9737	2.0120
		(.4426)	(.3580)	(.4426)	(.3583)	(.4552)	(.3739)	(.4550)	(.3769)
	50	3.0420	2.0067	3.0421	2.0069	2.9734	1.9676	2.9735	1.9738
		(.3076)	(.2540)	(.3076)	(.2543)	(.2836)	(.1904)	(.2837)	(.1910)
2	10	2.9352	1.9226	2.9352	1.9226	2.9203	1.9471	2.9202	1.9471
		(.6260)	(.4788)	(.6260)	(.4788)	(.6543)	(.4747)	(.6540)	(.4747)
	20	2.9869	1.9429	2.9869	1.9429	2.9802	1.9492	2.9801	1.9492
		(.4435)	(.3453)	(.4435)	(.3453)	(.4629)	(.3158)	(.4631)	(.3158)
	50	3.0014	1.9987	3.0014	1.9987	2.9671	2.0206	2.9668	2.0206
		(.2622)	(.2254)	(.2622)	(.2254)	(.2932)	(.1909)	(.2932)	(.1909)
2.25	10	3.1134	1.9122	3.1134	1.9102	2.9456	1.9938	2.9455	1.9707
		(.6496)	(.4488)	(.6496)	(.4476)	(.6050)	(.4935)	(.6051)	(.4878)
	20	3.0028	1.9623	3.0029	1.9620	2.9994	2.0074	2.9994	1.9964
		(.3995)	(.3057)	(.3996)	(.3056)	(.3601)	(.3131)	(.3601)	(.3114)
	50	2.9908	1.9774	2.9908	1.9773	2.9741	1.9893	2.9742	1.9850
		(.2841)	(.1738)	(.2841)	(.1739)	(.2933)	(.2063)	(.2929)	(.2059)
2.75	10	3.0725	1.9358	3.0842	1.9309	3.0003	1.8633	3.0003	1.8149
		(.5466)	(.3779)	(.5373)	(.3791)	(.5563)	(.3997)	(.5563)	(.3893)
	20	3.0437	1.9551	3.0437	1.9552	3.0552	1.9751	3.0552	1.9507
		(.4195)	(.3009)	(.4195)	(.2996)	(.3797)	(.2765)	(.3797)	(.2731)
	50	2.9726	2.0271	2.9726	2.0273	3.0183	2.0233	3.0180	2.0136
		(.2231)	(.1853)	(.2231)	(.1853)	(.2255)	(.1646)	(.2256)	(.1638)
3.25	10	2.9288	2.0104	2.9278	1.9990	3.0621	1.9547	3.0622	1.8874
		(.4649)	(.3823)	(.4655)	(.3816)	(.5005)	(.4018)	(.5004)	(.3879)
	20	3.0110	1.8980	3.0112	1.8977	2.9307	1.9773	2.9305	1.9450
		(.3331)	(.2656)	(.3331)	(.2647)	(.3920)	(.2538)	(.3921)	(.2497)
	50	3.0093	1.9245	3.0093	1.9239	2.9443	1.9957	2.9442	1.9830
		(.2071)	(.1580)	(.2071)	(.1581)	(.2079)	(.1732)	(.2080)	(.1721)

Table 2. Order-p normal distribution parameter estimates, all parameters unknown

	Maximum likelihood			Genetic algorithm			
	μ	σ_p	p	μ	σ_p	p	K
p=1							
n=50	3.0193	2.1429	1.2462	3.0267	2.1341	1.2250	
	(.2868)	(.3198)	(.3390)	(.2796)	(.3171)	(.3279)	10
n=100	2.9781	2.0606	1.1144	3.0074	2.0403	1.1099	
	(.2004)	(.1917)	(.1602)	(.2169)	(.1988)	(.1762)	5
p=1.25							
n=50	2.9765	2.0486	1.5072	3.0235	2.0528	1.4813	
	(.3578)	(.3436)	(.7016)	(.3176)	(.3237)	(.6211)	5
n=100	3.0447	2.0468	1.3691	3.0066	2.0593	1.3774	
	(.2181)	(.2495)	(.2954)	(.2063)	(.2820)	(.3766)	6
p=1.50							
n=50	2.9694	1.9805	1.7014	2.9812	2.0096	1.8719	
	(.3106)	(.3483)	(.7612)	(.3167)	(.3783)	(.9849)	2
n=100	2.9924	1.9949	1.5935	2.9922	2.0094	1.5980	
	(.1959)	(.2634)	(.3806)	(.2009)	(.2324)	(.4399)	1
p=1.75							
n=50	3.0454	2.1079	2.2741	2.9633	2.0760	2.1561	
	(.3220)	(.4431)	(1.0201)	(.2962)	(.3908)	(.9725)	2
n=100	2.9570	1.9861	1.8479	2.9894	1.9606	1.8190	
	(.1987)	(.2150)	(.3964)	(.2111)	(.2353)	(.4464)	
p=2							
n=50	3.0064	2.0554	2.5580	3.0510	2.0858	2.5986	
	(.2908)	(.3666)	(1.1405)	(.2963)	(.3090)	(1.0214)	1
n=100	2.9938	2.0654	2.2843	3.0127	2.024	2.2229	
	(.2196)	(.2362)	(.7219)	(.1815)	(.2246)	(.6650)	
p=2.25							
n=50	3.0160	2.0807	3.0267	2.9586	2.0637	2.8037	
	(.2985)	(.3049)	(1.1726)	(.2848)	(.3257)	(1.1711)	
n=100	3.0314	1.9904	2.4970	2.9944	2.0054	2.4778	
	(.2077)	(.2345)	(.6559)	(.1883)	(.2125)	(.7259)	
p=2.5							
n=50	2.9395	2.0682	3.2349	2.9985	1.9875	3.0510	
	(.2562)	(.2919)	(1.1717)	(.2670)	(.3054)	(1.2151)	
n=100	3.0771	2.0807	2.9473	3.0098	2.0395	2.7787	
	(.1718)	(.2123)	(.8330)	(.1661)	(.2171)	(.7440)	
p=2.75							
n=50	2.9752	2.0459	3.2150	3.0213	1.9974	3.2069	
	(.2416)	(.3026)	(1.1268)	(.2671)	(.3146)	(1.1257)	
n=100	2.9778	2.0149	3.0505	2.9904	1.9498	2.8491	
	(.1846)	(.2213)	(.8972)	(.1846)	(.1966)	(.7361)	
p=3							
n=50	2.9970	2.0584	3.6761	3.0009	2.0152	3.5714	
	(.2448)	(.2661)	(1.1710)	(.2509)	(.3022)	(1.2258)	
n=100	3.0188	2.0110	3.4036	3.0051	2.0365	3.3715	
	(.1733)	(.2124)	(.9240)	(.1624)	(.1970)	(.8667)	
p=3.25							
n=50	3.0054	1.9841	4.0412	3.0103	1.9782	3.7364	
	(.2199)	(.2248)	(1.0605)	(.2260)	(.2782)	(1.2226)	
n=100	2.9912	2.0242	3.6140	3.0063	2.0174	3.5910	
	(.1609)	(.1932)	(.9203)	(.1639)	(.2161)	(.9309)	

mean and standard errors of the parameter estimates was made by drawing 100 samples using of size $n = 10, 20, 50$, when p was supposed known, and $n = 50, 100$, when p was unknown, from a normal distribution of order p ($\mu = 3; \sigma_p = 2$) with $p = 1, 1.25, 1.5, 1.75, 2, 2.25, 2.75$ and 3.25. When p was assumed known, the ML, jackknife, GA and SA estimators were considered for estimating the location and scale parameters μ and σ_p. Though not clearcut results were obtained, in most cases the GA seemed able to ensure better performances than the other procedures with respect to both the bias and the standard errors of the estimates. For the p unknown case, the GA yields estimates of p less biased than the ML approach when $p > 2$. Opposite results occur if $p \leq 2$. In this case, however, some samples were discarded by the ML procedure. Similar findings were obtained, except some particular cases, for the estimates of the two other parameters.

References

AGRO', G. (1995): Maximum Likelihood Estimation for the Exponential Power Function Parameters. *Communications in Statistics – Simulation and Computation, 24(2)*, 523–536.

AGRO', G. (1999): Parameter Orthogonality and Conditional Profile Likelihood: the Exponential Power Function Case. *Communications in Statistics – Theory and Methods, 28(8)*, 1759–1768.

BOX, G.E.P. (1953): A Note on Regions for Test and Kurtosis. *Biometrika, 40*, 465–468.

BOX,G.E.P and TIAO, G.C. (1973): *Bayesian Inference in Statistical Analysis.* Addison-Wesley, Reading, Massachussetts.

BROOKS, S.P. and MORGAN, B.J.T. (1995): Optimization Using Simulated Annealing, *The Statistician, 44*, 241–257.

BURGIO, G. and NIKITIN, Y. (1998): Goodness-of-fit Tests for Normal Distributions of Order p and their Asymptotic Efficiency. *Statistica, 58*, 213–230.

CHATTERJEE, S. and LAUDATTO, M. (1997): Genetic Algorithms in Statistics: Procedures and Applications. *Communications in Statistics – Simulation and Computation, 26(4)*, 1617–1630.

CHIODI, M. (1986): Procedures for Generating Pseudo-random Numbers from a Normal Distribution of Order p (p> 1), *Statistica Applicata, 1*, 7–26.

CHIODI, M. (1988): *Sulle Distribuzioni di Campionamento delle Stime di Massima Verosimiglianza dei Parametri delle Curve Normali di Ordine p.* Istituto di Statistica Università di Palermo, Palermo.

CHIODI, M. (1994): Approssimazioni Saddlepoint alle Distribuzioni Campionarie degli Stimatori di Massima Verosimiglianza dei Parametri delle Curve Normali di Ordine p per Piccoli Campioni. *Atti della XXXVII Riunione Scientifica della SIS (Italian Statistical Society), 1*, 139–146.

CHIODI, M. (2000): Le Curve Normali di Ordine p nell'Ambito delle Distribuzioni di Errori Accidentali: Una Rassegna dei Risultati e Problemi Aperti per il Caso Univariato e per quello Multivariato. *Atti della XL Riunione Scientifica della SIS (Italian Statistical Society)*, 59–60. Extended and revised version available at http://dssm.unipa.it/chiodi/

DIANANDA, P.H. (1949): Note on Some Properties of Maximum Likelihood Estimates. *Proceedings of the Cambridge Philosophical Society, 45*, 536–544.

HOLLAND, J.H. (1975): *Adaptation in Natural and Artificial Systems.* University of Michigan Press, Ann Arbor, Michigan. Second Edition: MIT Press, 1992.

LA TONA, L. (1994): Uno Stimatore Jackknife del Parametro di Scala di una Curva Normale di Norma p. *Metron, 52*, 181–188.

LUNETTA, G. (1963): Di una Generalizzazione dello Schema della Curva Normale. *Annali della Facoltà di Economia e Commercio di Palermo, 17, 2*, 237–244.

LUNETTA, G. (1966): Di alcune Distribuzioni Deducibili da una Generalizzazione dello Schema della Curva Normale. *Annali della Facoltà di Economia e Commercio di Palermo, 20, 1*, 117–143.

MINEO, A. (1989): The Norm-p Estimation of Location, Scale and Simple Regression Parameters. *Lecture Notes in Statistics, Statistical Modeling Proceedings.* Springer, Heidelberg, 222–223.

MINEO, A.M. (1994): Un Nuovo Metodo di Stima di p per una Corretta Valutazione dei Parametri di Intensità e di Scala di una Curva Normale di Ordine p. *Atti della XXXVII Riunione Scientifica della SIS (Italian Statistical Society), 2*, 147–154.

SUBBOTIN, M.T. (1923): On the Law of Frequency of Errors. *Matematicheskii Sbornik, 31*, 296–301.

TURNER, M.E. (1960): On Heuristic Estimation Methods. *Biometrics, 16, 2*, 299–302.

VIANELLI, S. (1963): La Misura della Variabilità Condizionata in uno Schema Generale delle Curve Normali di Frequenza. *Statistica, 33*, 447–474.

Part IV

Classification and Data Analysis in Economics

Non-linear Dynamics in the Industrial Production Index*

Alessandra Amendola and Giuseppe Storti

Dipartimento di Scienze Economiche e Statistiche, Università degli Studi di Salerno, Via Ponte Don Melillo, 84084, Fisciano(SA), Italy.

Abstract. The paper presents an approach to modelling and forecasting the Industrial Production Index which allows to account for the presence of asymmetric effects in both the conditional mean and the conditional variance. More precisely, the proposed approach combines a Self Exciting Threshold AutoRegressive (SETAR) model for the conditional mean with a conditional heteroskedastic model fitted to the residuals. Namely, we use a Constrained Changing Parameters Volatility (CPV-C) model which allows to capture asymmetries in the conditional variance dynamics by means of interaction terms between past shocks and volatilities. The out of sample fitting performance of the model is evaluated by means of an application to a time series of U.S. data.

1 Introduction

In recent years, there has been a great interest in modelling non-linearity in economic time series (Potter (1995); Clements and Krolzig (1998)). In the Industrial Production Index, as in a wide range of economic time series, the main source of non-linearity is given by the asymmetric nature of the business cycle. Furthermore, evidence of conditional heteroskedasticity in the Industrial Production Index series has been found by some authors (e.g. Byers and Peel (1995)). The approach we propose in this paper allows to have asymmetric effects present not only in the level but also in the conditional variance of the series. The main intuition behind this is that a negative shock, corresponding to a decrease in the industrial production level, is expected to have an effect on the volatility of the series which is greater than that due to a positive shock of the same magnitude, corresponding to an increase in the industrial production level. In particular we combine a threshold

* This paper was supported by the MURST COFIN 2000 project: "Modelli Stocastici e Metodi di Simulazione per l'Analisi di Dati Dipendenti".

model for the conditional mean with a Constrained Changing Parameters Volatility (CPV-C) model, recently proposed by Storti (1999). The out of sample fitting performance of the model is assessed by means of some well known measures of goodness of fit. The results are compared to those obtained by a model including a simple linear autoregressive component for the conditional mean together with a CPV-C model for the conditional variance of the series. The paper is organised as follows: section 2 describes the modelling strategy; the results of an application to the U.S. seasonal adjusted Industrial Production Index series are shown in section 3. Some concluding remarks are given in section 4.

2 Modelling Asymmetry

2.1 Asymmetry in the TAR model

In many economic time series the non-linearity is typically due to the presence of different behaviours in negative and positive growth phases. Therefore, the response of the output to shocks occurring at different stages of the business cycle is asymmetric. This asymmetry in the level can be captured by a threshold model in which the different states are represented by the different regimes. In particular we consider the class of Threshold AutoRegressive (TAR) models first presented by Tong (1978) and further developed and applied in Tong and Lim (1980) (a more thorough discussion can be found in Tong (1990)). A TAR model can be regarded as a piecewise linear autoregressive structure, which allows to obtain the decomposition of a complex stochastic system into smaller subsystems, based on the values assumed by a *threshold variable* compared with a set of predetermined *threshold values.* Let $\{Y_t\}$ be an observed time series, a TAR model for Y_t is given by:

$$\begin{aligned} & Y_t = a_0^{(j)} + \sum_{i=1}^{p} a_i^{(j)} Y_{t-i} + u_t \\ & X_{t-d} \in R_j \end{aligned} \tag{1}$$

where $\{u_t\}$ is i.i.d. with zero mean and finite variance, the threshold values $\{r_0, r_1, r_2, .., r_s\}$, are such that $r_0 < r_1 < ... < r_s$,

$r_0 = -\infty$ and $r_s = +\infty$ with $R_j = (r_{j-1}, r_j]$, d is a positive integer. If we choose as a threshold variable a past realisation of Y_t, $\{Y_{t-d}\}$, the time series $\{Y_t\}$ follows a Self Exciting Threshold AutoRegressive (SETAR) model. To investigate the appropriateness of a threshold non-linear model instead of a simpler linear model, we perform the Tsay test for detecting threshold non-linearity presented in Tsay (1989) and successively refined and generalised by Tsay (1998). Under the null hypothesis of a linear model, Tsay's test is asymptotically distributed as a chi-squared random variable with (p+1) degrees of freedom.

2.2 Asymmetric effects in the CPV model

The Constrained Changing Parameters Volatility (CPV-C) model, discussed in Storti (1999) and Amendola and Storti (2002), allows to model asymmetries in the conditional variance by considering interactions between past shocks and volatilities. Let $\{u_t\}$ be an observed time series such that $(u_t|I^{t-1}) \sim (0, h_t^2)$, $h_t^2 < \infty$, $\forall t$, with I^{t-1} indicating the set of information available at time (t-1). A CPV-C model of order (m, n), is defined as:

$$u_t = \sum_{i=1}^{m} a_{i,t} u_{t-i} + \sum_{j=1}^{n} b_{j,t} h_{t-j} + e_t \tag{2}$$

where $a_{i,t}$ and $b_{j,t}$ are sequences of iid random variables and, furthermore, given $\theta_{\mathrm{t}} = [a_{1,t}, ..., a_{m,t}|b_{1,t}, ..., b_{n,t}]'$:

$$E(\theta_t) = E(\theta_t|I^{t-1}) = \mathbf{0} \qquad Var(\theta_t) = Var(\theta_t|I^{t-1}) = \mathbf{Q}$$

for $i = 1,, m$; $j = 1,, n$ and $t = 1,, T$. Also, let $e_t \sim N(0, \omega)$ be a Gaussian white noise error independent of θ_t. More details on the relationship between random coefficient autoregressive models and conditional heteroskedastic models can be found in Tsay (1987). For a CPV-C model of order (m, n), the conditional variance of u_t given past information I^{t-1} is given by:

$$h_t^2 = \omega + \sum_{j=1}^{m} \alpha_j u_{t-j}^2 + \sum_{j=1}^{n} \beta_j h_{t-j}^2 + 2 \sum_{i=1}^{m} \sum_{j=1}^{n} \gamma_{i,j} u_{t-i} h_{t-j} + \tag{3}$$
$$+ 2 \sum_{i \neq j} \delta_{i,j} u_{t-i} u_{t-j} + 2 \sum_{i \neq j} \eta_{i,j} h_{t-i} h_{t-j}$$

where $\{\omega, \alpha_j, \beta_j, \gamma_{i,j}, \delta_{i,j}, \eta_{i,j}\}$, for $i = 1, ..., m$ and $j = 1, ..., n$, are unknown parameters which can be estimated by maximum likelihood. Our experience suggests that economic and financial data currently encountered in practical applications do not usually require model orders $m, n > 2$ and that the simple CPV-C (1,1) offers a flexible framework for the estimation of h_t^2 in a wide range of situations. In this case the equation for the conditional variance simplifies to:

$$h_t^2 = \omega + \alpha_1 u_{t-1}^2 + \beta_1 h_{t-1}^2 + 2\gamma_{1,1} u_{t-1} h_{t-1} \tag{4}$$

where the asymmetry in the relation between the conditional variance and past shocks comes from the interaction term $2\gamma_{1,1} u_{t-1} h_{t-1}$ In particular, for $\gamma_{1,1} < 0$, a positive penalty term will be added to h_t^2 if $u_{t-1} < 0$ while a positive quantity will be subtracted from h_t^2 if $u_{t-1} > 0$. It is also worth noting how the magnitude of the penalty term added to the conditional variance does not depend only on the magnitude of the shock itself but this is weighted or, more properly, rescaled by the conditional standard deviation at time (t-1). So the greater will be the uncertainty associated with the negative shock u_{t-1}, the greater will be the impact of u_{t-1} on h_t^2.
A similar reasoning applies to the case in which $\gamma_{1,1} < 0$. A generalization which allows for simultaneous modelling of conditional mean and variance is given by the regression CPV-C model

$$y_t = \mathbf{M}_t \beta + u_t \tag{5}$$

where $\mathbf{M}_t$ is a vector $(1 \times g)$ of regressors, β a vector $(g \times 1)$ of unknown parameters and u_t follows a CPV-C model. The formulation of model (5) is quite general and the regression term included into the observation equation can incorporate an ARMA typc structure with exogenous explanatory variables. Also, replacing the constant parameter vector β by a time variable vector β_t, we can easily accommodate for some common non-linear structures such as Threshold Autoregressive models. The model parameters can be estimated maximizing a Gaussian log-likelihood function expressed in the classical prediction error decomposition form. Under the above assumptions, the log-likelihood function of a regression CPV-C model can be written as

$$\ell(y; \beta_t, \omega, \mathbf{Q}) = -\frac{T}{2}\log(2\pi) - \frac{1}{2}\sum_{t=1}^{T}\log\left[(h_t)^2\right] - \frac{1}{2}\sum_{t=1}^{T}\frac{(e_{t|t-1})^2}{(h_t)^2} \tag{6}$$

where $e_{t|t-1} = y_t - \mathbf{M}_t\beta_t$. The unknown parameters in $\{y; \beta_t, \omega, \mathbf{Q}\}$ can be estimated by maximising the Gaussian log-likelihood (6) using a version of the EM algorithm tailored for state space models by Wu et al. (1996). Alternatively scoring or quasi-Newton methods could also be used (see Watson and Engle (1983), for a discussion on the application of the method of scoring in the context of state space models). The asymptotic standard errors can be obtained by analytical evaluation of the Observed Information matrix. More details can be found in Storti and Vitale (2002).

3 The U.S. Industrial Production Index

The dataset we analyze is given by the series of the monthly seasonal adjusted U.S. Industrial Production Index, from January 1940 to December 2000, for a total of 732 observations. The first 696 observations have been used for model identification and estimation while the last 36 have been used to assess the out of sample fitting performance. The analysis is performed on the growth rate series calculated as the logarithmic first difference $Y_t = \nabla(\log X_t)$. The time plots of the original and transformed series have been reproduced in Figure 1. As a first step, on the basis of the estimated autocorrelation function, we identify and estimate a linear autoregressive model of order 1 with a CPV-C(1,1) structure for the conditional variance[1]. This will be used as a benchmark model for our analysis. The estimated parameters with their respective standard errors are reported in Table 1. An important point is that the estimate obtained for the $\gamma_{1,1}$ parameter is negative and significantly different from 0, showing evidence of asymmetry in the conditional variance dynamics. In order to investigate the presence of threshold non-linearity we use Tsay's linearity test (see Tsay (1998)). The results shown in Table 2 lead to reject the null hypothesis of linearity. Furthermore, the test suggests using a delay $d = 4$.

Table 1. Parameter estimates and asymptotic s. e. of the AR-CPV.

AR(1)	a_0	a_1		
	0.0034 (0.0008)	0.3916 (0.0348)		
CPV-C(1,1)	ω	α_1	$\gamma_{1,1}$	β_1
	2.4189e-5 (4.2081e-6)	0.2892 (0.0542)	-0.1888 (0.0310)	0.5371 (0.0581)

Table 2. Results of Tsay's threshold non-linearity test.

d	1	2	3	4	5	6	7
Test	2.097	17.91	25.13	**29.90**	6.935	10.25	16.93
d.f.	2	2	2	**2**	2	2	2

Setting the AR order to lie in the range [1,2], given d=4 and the number of regimes $s \in [2,3]$, among all the possible combinations of p, d and s we have then chosen the model characterised by the minimum Akaike Information Criterion (AIC) and selected the threshold values $\hat{r}_1$=-0.0018 and $\hat{r}_2$=0.0065 giving the mini-

[1] The decision to include a CPV-C conditional heteroskedastic component is based on the results of an ARCH-LM test performed on the residuals obtained from fitting an autoregressive model of order 1 to the data. The value of the test statistic suggests to neatly reject the null hypothesis of a simple linear model in favour of the alternative of conditional heteroskedasticity. These results are available upon request.

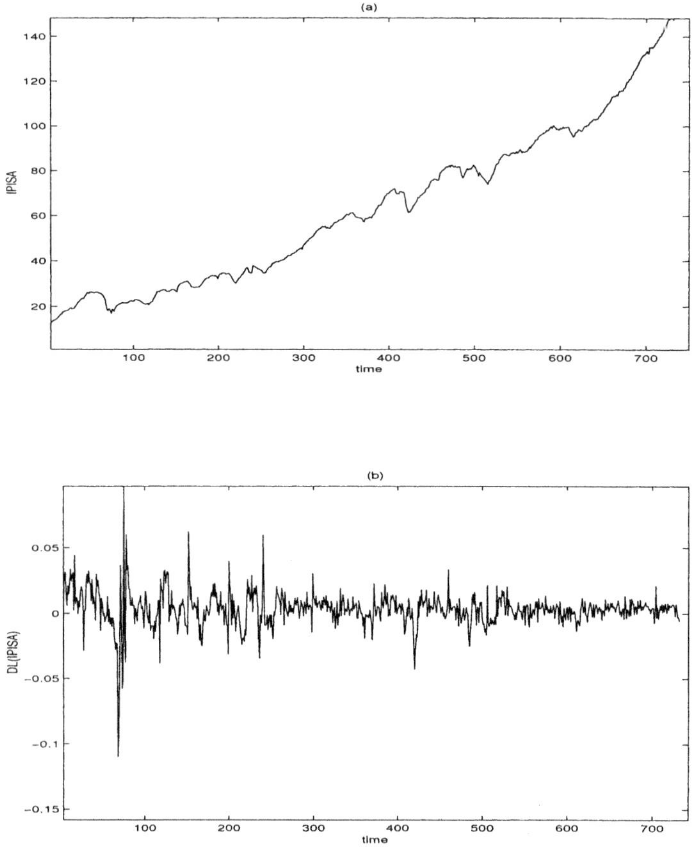

Fig. 1. Original data (a) X_t and first difference of log-transformed data (b) $Y_t = \nabla(\log X_t)$, for the U.S. seasonal adjusted (SA) IPI series from January 1940 to December 2000.

mum AIC. The least squares estimates of the parameters of the final model, a SETAR(3;1,1,1), and the number of observations in each regime are given in Table 3. The results of a Lagrange Multiplier test for AutoRegressive Conditional Heteroskedasticity (ARCH-LM; Engle (1982)) performed on the residuals of the estimated SETAR model (Table 4) and the analysis of the autocorrelation function of the squared residuals suggest the presence of conditional heteroskedasticity. Hence, we estimate a CPV-C(1,1) model for the conditional variance. The maximum likelihood estimates of the model parameters and their asymptotic standard errors are reported in Table 5. Again the negative sign of $\gamma_{1,1}$

Table 3. Least squares estimates and s. e. for the SETAR-CPV(3;1,1,1) model.

Regime	n.obs.	a_0	a_1
I	186		0.4967 (0.0678)
II	270	0.0009 (0.0005)	0.3678 (0.0481)
III	235	0.0064 (0.0009)	0.1263 (0.0656)

Table 4. Results of the ARCH LM test up to lag 3.

ARCH-LM Test			
Test	105.5662	**p-value**	0.0000

shows evidence of asymmetry in the conditional variance of the series. This is confirmed by the news impact curve associated to the estimated model and reported in Figure 2.

The out-of-sample fitting performance of the SETAR-CPV model

Table 5. ML parameter estimates and asymptotic s. e. for the estimated CPV-C(1,1) model.

ω	α_1	$\gamma_{1,1}$	β_1
1.0158e-5 (2.1726e-6)	0.2354 (0.0420)	-0.1401 (0.0247)	0.7109 (0.0368)

is assessed by means of some widely used measures of goodness of fit (MSE, MAE, MSPE, RMSPE, MAPE) and the results are compared with those obtained by the AR-CPV model (Table 6). On the basis of all the indices considered, the one-step-ahead predictions obtained by the SETAR-CPV model are superior to those generated by the AR-CPV model.

Furthermore, from an economic standpoint, a switching regime model is more appealing since it identifies different models to characterize the different states of the system. Hence, the final model structure is consistent with the widely shared assumption that, during recessions, the industrial production evolves accord-

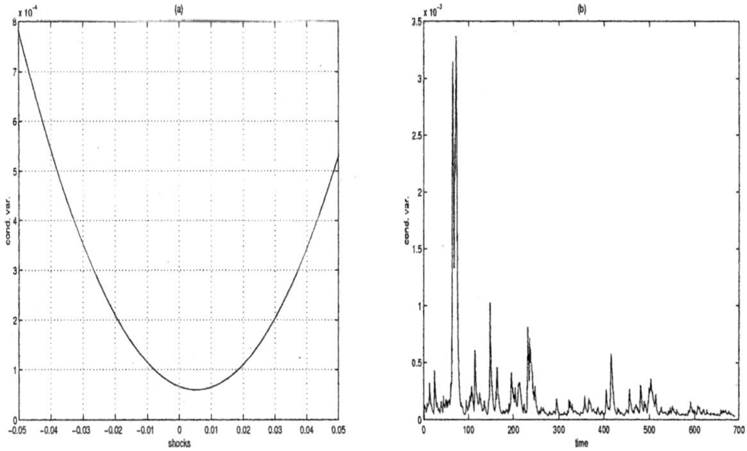

Fig. 2. News impact curve for the estimated CPV-C(1,1) model (a) and estimated conditional variance (b).

ing to a dynamic pattern substantially different from that followed during growth phases.

The one-month-ahead predictions for the period going from January 1998 to December 2000 have been represented in Figure 3 together with the estimated approximate confidence bands placed at ± 2 and ± 3 conditional standard deviations.

Table 6. Out of sample fitting performance of the SETAR-CPV compared to the AR-CPV.

	MSE $\times 10^4$	MAE	MSPE	RMSPE	MAPE
SETAR-CPV	0.31745	0.00392	7.68511	2.77220	1.46072
AR-CPV	0.32599	0.00408	16.56569	4.07009	1.84026

4 Concluding Remarks

A novel approach to modelling asymmetry in the conditional mean and variance of the Industrial Production Index has been proposed. The overall performance of the proposed model has to be considered satisfactory. Our approach allows to detect and estimate the impact of asymmetry on both the conditional mean

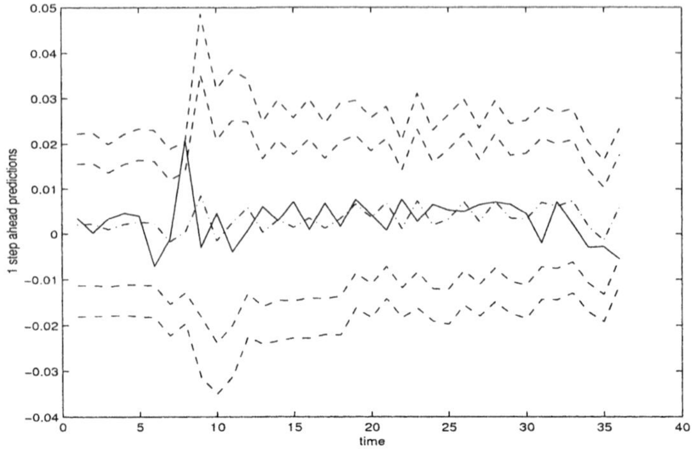

Fig. 3. Observed data (solid) and one-month-ahead forecasts (dashed) with ±2, ±3 conditional s.d. approximate confidence bands for the estimated SETAR-CPV model.

and variance of the series. Since asymmetry is introduced into the model for the conditional variance via interactions between past observations and volatilities, the effect of a negative shock on the current volatility depends not only on the magnitude of the shock but also on economy's uncertainty conditions as measured in terms of past volatilities. In order to further investigate the non-linear structure of the Industrial Production Index series we believe that in our future work it will be interesting to compare the results here presented with those obtained analysing a wide range of series from other countries and, at the same time, to consider other non-linear methods. Useful indications could also be gained by looking at longer forecast horizons.

References

AMENDOLA, A. and STORTI, G. (2002): A Non-Linear Time Series Approach to Modelling Asymmetry in Stock Market Indexes, *Statistical Methods and Applications*, *11*, *2*, 201–216.

BYERS, J.D. and PEEL, D.A. (1995): Forecasting Industrial Production Using Non-linear Methods, *Journal of Forecasting*, *14*, 325–336.

CLEMENTS, M.P. and KROLZIG, H.M. (1998): A comparison of the forecast performance of Markov-switching and threshold autoregressive models of US GNP, *Econometrics Journal*, *1*, C47–C75.

ENGLE, R.F. (1982): Autoregressive conditional heteroskedasticity with estimates of the variance of the United Kingdom inflation, *Econometrica*, *50*, 987–1007.
POTTER, S.M. (1995): A Nonlinear Approach to US GNP, *Journal of Applied Econometrics*, *10*, 109–125.
STORTI, G. (1999): The CPV model: a state space generalization of GARCH processes, *Proceedings of the conference SCO 99: 'Modelli Complessi e Metodi Computazionali Intensivi per la Stima e la Previsione'*, 248–253.
STORTI, G. and VITALE, C. (2002): Likelihood inference in BL-GARCH models, submitted for publication to *Computational Statistics.*
TONG, H. (1978): On a Threshold Model, in C.H. Chen (Ed.): *Pattern Recognition and Signal Processing*, Sijthoff & Noordhoff, Amsterdam.
TONG, H. (1990): *Non-linear time series: a dynamical system approach*, Oxford University Press, Oxford.
TONG, H. and LIM, K.S. (1980): Threshold autoregression, limit cycles and cyclical data, *Journal of the Royal Statistical Society, B*, *42*, 245–292.
TSAY, R.S. (1987): Conditional heteroscedastic time series models, *Journal of the American Statistical Association*, *82*, 590–604.
TSAY, R.S. (1989): Testing and modelling threshold autoregressive processes, *Journal of the American Statistical Association*, *84*, 231–240.
TSAY, R.S. (1998): Testing and modelling multivariate threshold models, *Journal of the American Statistical Association*, *93*, 1188–1202.
WATSON, M.W. and ENGLE, R.F. (1983): Alternative algorithms for the estimation of dynamic factor, MIMIC and varying coefficient regression models, *Journal of Econometrics*, *23*, 385–400.
WU, L. S., PAI, J.S. and HOSKING, J.R.M. (1996): An algorithm for estimating parameters of state-space models, *Statistics & Probability Letters*, *28*, 99–106.

Tensorial Co-Structure Analysis for the Full Multi Modules Customer Satisfaction Evaluation

Pietro Amenta[1] and Pasquale Sarnacchiaro[2]

[1] Department of Economics, Mathematic and Statistics University of Lecce. E-mail: amenta@economia.unile.it
[2] Department of Mathematics and Aeronautics - Faculty of Engineering University of Naples "Federico II". E-mail: sarnacch@unina.it

Abstract. In literature several methods have been proposed to evaluate Customer Satisfaction (CS), often based on the gap concept. This paper sets a strategy of integrated analysis which in the presence of several observations for the same statistical unit, divides the gap between customer's quality expectation and quality perception into further elements to be analyzed in order to improve the service quality.

1 Introduction

During the last twenty years firms' strategy has gradually shifted from marketing to Total Quality Management to Customer Satisfaction. Particularly, for a company, the knowledge of the customer evaluation of a given service represents an important starting point for every business strategy. In literature several methods have been proposed for the service quality assessment. A lot of these models (Servqual, Normed Quality, Qualitometro etc.) are based on the Gap Theory of Service Quality, developed for the for-profit sector by Parasuraman et al. (1994). These models measure the gap between customer's expectations for excellence and their perceptions of actual service delivered, so service providers can understand both customer expectations and their perceptions of specific services. SERVQUAL undertakes to measure service quality across five dimensions: reliability, assurance, tangibles, empathy and responsiveness. The aggregation of these measures creates some problems on the immaterial subjective evaluations: a) we have to cope with immaterial subjective evaluations and with not fully comparable ordinal scale of measures which are treated as interval scale; b) the choice of an optimal multidimensional statistical analysis for these data. Several techniques have been proposed in literature for the description and the exploratory study of these data. Aims of these proposals are to estimate the multidimensional aspects of the investigated system and the introduction of criteria of judgment (Lauro et al. (1997); D'Ambra et al. (1999) [D&al]; D'Ambra, Amenta (2000a); D'Ambra, Amenta (2000b) [D&A]). A deeper analysis can be considered, in the student satisfaction, if we have questionnaires taken in different K modules or times (for example, during a post-degree training course with different teachers): we have K pre-service matrices (one for each module) and K post-service matrices (at the end of each module).

Aim of this paper is to propose a tensorial extension of Torre and Chessel's approach in order to analyse this complex expectations-perceptions system (Full Multi Modules).

2 Tensorial Co-Structure Analysis

In order to evaluate the student satisfaction, we consider a questionnaire, taken on n students, in different K modules during a training course with K different teachers: it leads to have K pre-service and K post-service matrices at the start and at the end of each module. In the assessment of the service performance, the used scale of measure is necessarily of ordinal type. This scale involves a problem of comparability and, moreover, several techniques of the multidimensional data analysis deal this scale as a quantitative one. To overcome this problem we could make a data pre-coding by using one of the most popular techniques proposed in literature in order to transform an ordinal scale to an interval one: the psychometric approach of Thurstone. Data are represented in terms of matrices X_k (pre-service) and Y_k (post-service), $(k = 1, \ldots, K)$, of order $(n \times p)$. Let Q be positive defined symmetrical matrix (metric) of dimensions $(p \times p)$ and $D_n = diag(d_1, \ldots, d_n)$, $(\sum_{i=1}^{n} d_i = 1 \ )$, weight matrix of statistical units. For each k we obtain couples of statistical triplets: (X_k, Q, D_n) and (Y_k, Q, D_n). Following Lafosse (1989), every couple can be defined *fully matched* because they describe the same statistical units (n rows) by means the same variables (p columns). Let $\tilde{X}$ and $\tilde{Y}$ be three way arrays, of order $(n \times p \times k)$, which collect these matrices X_k and Y_k, respectively. $\tilde{X}$ and $\tilde{Y}$ can be considered, also, *fully matched array.* We transpose arrays $\tilde{X}$ and $\tilde{Y}$, by stopping the first dimension, so to have $\tilde{X}'$ and $\tilde{Y}'$ of order $(n \times k \times p)$. We vectorialise, then, the two-way matrices X_j of $\tilde{X}'$ and Y_j of $\tilde{Y}'$, along the third dimension j $(j = 1, \ldots, p)$: ${}_jx = Vec(X_j)$ and ${}_jy = Vec(Y_j)$, both belonging to the vectorial space $\Re^n \otimes \Re^k$; we collect these vectors, by columns, in the matrices $X = [{}_1x, ..., {}_px]$ (see also in Dynamic Factorial Analysis, (Coppi, Zannella, 1979)) and $Y = [{}_1y, ..., {}_py]$ of order $(nK \times p)$, respectively, and let, also, be $M = D_n \otimes I$ with I identity matrix. We obtain fully matched statistical triplets yet: (X, Q, M) and (Y, Q, M). Starting from this vectorialisation and the Co-Structure analysis (Torre, Chessel, 1995) [T&C], which can be considered as a particular case of the Co-inertia analysis (Chessel, Mercier, 1993 [C&M]; Tucker, 1958), we search the co-structure axis which maximises the covariance between the coordinates of the row projections of the two matrices on this axis. The Tensorial Co-Structure Analysis maximises the following amount:

$$\begin{cases} \max \langle Xa, Ya \rangle \\ \|a\|_Q = 1 \end{cases} \text{ or, equivalently, } \begin{cases} \max \sum_k^K \langle X_k Qa, Y_k Qa \rangle \\ \|a\|_Q = 1 \end{cases}$$

where a belongs to $\Re^p$. This methodology simultaneously carries out the analysis of inertia of each matrix and it considers the maximum of covariance between the coordinates of two clouds on vector a (Tab.1). The maximum is reached for the first eigenvector of the matrix $0.5 * Q^{0.5}(X'MY + Y'MX)Q^{0.5}$.

Table 1: Co-Inertia and Co-Structure of fully matched tables criteria

[C&M,93]	[T&C,95][D&al,99]	[D&A,2000b]	Staticos
$\langle X_1 Qa, Y_1 Qb \rangle_{D_n}$	$\langle X_1 Qa, Y_1 Qa \rangle_{D_n}$ Single Module	$\sum_{k=1}^{K} \langle X_1 Qa, Y_k Qa \rangle_{D_n}$ Multi Modules	$\sum_{k=1}^{K} \langle X_k Qa, Y_k Qa \rangle_{D_n}$ Full Multi Modules
Co-Inertia	*Co-Structure*		
X_1 and Y_1 are referred to the expectation and perception case single module			

The complete matching of the two matrices X and Y leads to consider the statistical units of the two tables as elements of a single space and to develop, according to the gap theory, the analysis of the triplet $(X-Y,Q,M)$. While the co-structure analysis represents the resemblance core between X and Y, the analysis of triplet $(X-Y,Q,M)$ highlights in what they are different. This leads to consider the eigenvectors associated to the decomposition of the operators $(X-Y)'M(X-Y)Q$ linked to the maximum eigenvalue λ. Let In_X, In_Y and In_D be the inertias associated to the analysis of the triplets (X,Q,M), (Y,Q,M) and $(X-Y,Q,M)$, respectively. The Co-Structure and Difference Analysis of fully matched tables are linked by the relationship $In_D = In_X + In_Y - 2tr(X'MYQ)$; it decomposes the inertia of the difference matrix into three components and it allows: to analyse the customer satisfaction evaluation based on the gap, to consider the single pre-service and post service matrices and their link. The co-structure and different analysis can be performed, also, if we vectorialise the two-way X_k matrices $(n \times p)$ of $\tilde{X}$ and Y_k of $\tilde{Y}$, along the third dimension k: ${}_kx = Vec(X_k)$ and ${}_ky = Vec(Y_k) \in \Re^n \otimes \Re^p$. Let $\hat{X} = [{}_1x, ..., {}_Kx]$ and $\hat{Y} = [{}_1y, ..., {}_Ky]$ be matrices of order $(np \times k)$, respectively, and let, also, be $\hat{M} = D_n \otimes Q$. We obtain the fully matched statistical triplets: $(\hat{X}, I, \hat{M})$ and $(\hat{Y}, I, \hat{M})$ as well as the relationship $In_{\hat{D}} = In_{\hat{Y}} + In_{\hat{X}} - 2tr(\hat{X}'\hat{M}\hat{Y})$ associated to the analysis of the triplets $(\hat{X}, I, \hat{M})$, $(\hat{Y}, I, \hat{M})$ and $(\hat{X}-\hat{Y}, I, \hat{M})$, respectively. For this approach, performed on (X,Y) and/or $(\hat{X},\hat{Y})$, we have the following properties or particular cases:

- The analysis of the triplets (X,Q,M) $[(\hat{X}, I, \hat{M})]$ and (Y,Q,M) $[(\hat{Y}, I, \hat{M})\,]$ represent the Interstructure phase of Statis (Lavit, 1988) developed on the pre-service matrices X_j $[X_k]$ and the post-service matrices Y_j $[Y_k]$, respectively;
- Similarly, the analysis of the triplet (X,Q,M) $[(Y,Q,M)]$ is the same, also, to achieve $\max a'(\sum_{k=1}^{K} X_k' D_n X_k Q)a$ $[\max a'(\sum_{k=1}^{K} Y_k' D_n Y_k Q)a]$ with the constraint $a'Qa = 1$;
- The analysis of the triplet $(\hat{Y}, I, \hat{M})$ could be also considered as a *Multi-Modules* version of SERVPERF (Cronin, Taylor, 1992) with unitary weights for the dimensions and equal number of items for each dimension;

- The analysis of $X'MYQ$ is the Tensorial Co-Structure Analysis, which can be considered, also, as a Co-Structure approach of STATICO (Hanafi, 1998) that we call STATICOS (for Statis and Co-Structure);
- The analysis of the triplet $(\hat{X} - \hat{Y}, I, \hat{M})$ represents the Inter-structure phase of Statis developed on the gap-matrices $X_k - Y_k$ (see also Lauro et al. (1997));
- the Tensorial Co-Structure Analysis, performed on (X, Y), can be considered a generalization of the proposals of D'Ambra et al. (1999) and D'Ambra, Amenta (2000b), which are particular cases (Tab.1).

3 An Example of "Full Multi-Modules" Customer Satisfaction Evaluation

Usually the services can be distinguished in four typologies: 1) Interface service, the supply provides an interaction between supplier and client/user (ex. Training service), 2) Data processing service, the supply provides a real good (e.g. Word processor, Research program), 3) Availability service, it gives several objects at someone's disposal (ex. To rent a car), 4) Process service, the good availability is guaranteed by a process control (ex. Electric power supply). In this example we consider the assessment of an interface service: the student satisfaction evaluation. Regarding to the method to use for the survey, we have two alternatives: the first one (direct method) involves the student in the process of evaluation, the second one (indirect method) uses both information given by the personnel of contact through their relationship with the student and possible claims. We choose the first method and we use a questionnaire, to administer to the students, in order to have the necessary information for the assessment. The questionnaire takes into account all the characteristics that influence the student satisfaction. These characteristics have been organised in: Teaching and General Organisation Areas.

In order to minimise the level of intrusion in the survey, the expectations and the perceptions of the students are taken at the starting and at the end of each module. The questionnaire has

been given to 69 students. We have:
- 3 expectation matrices ("Exp Mod1", "Exp Mod2" and "Exp Mod3")
- 3 perception matrices ("Perc Mod1", "Perc Mod2" and "Perc Mod3").
The 20 questions are about: Accessibility, Parking, Rooms, Teaching tools, Timetable, Length of course, Course supply capability, General Information, Technical Service, Nursing, (General Organisation group); Teaching books utility, Teaching equipment clearness, Lessons, Teacher availability, Practical work, Acquired ability, Teacher ability, Teacher competence, Teacher involvement, Teacher kindness (Teaching group). Due to the nature of the variable (expressed on ordinal scale) the raw scores have been transformed by Thurstone normal model in order to substitute the original ordinal data with quantitative data. We develop both a Tensorial Co-structure analysis (STATICOS) and a Difference analysis of "Multi-Modules" Fully Matched Tables regarding the student satisfaction evaluation of the course.

A co-structure test (row permutations) has been performed on the data obtaining a significative value for the co-structure between pre and post-service matrices.

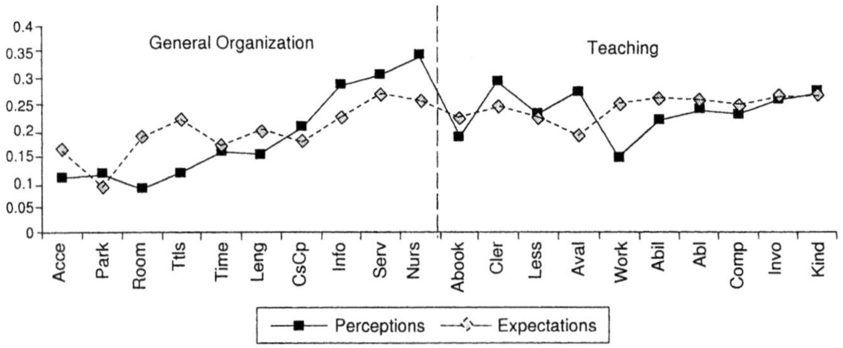

Fig. 1. The first axis of Staticos.

In Figure 1 we have the projections of the global scores of the expectation and perception characteristics on the first axis of Staticos. It seems that, in the second area, the values of perceptions often exceed the expectations, with a neutral opposition in the first area: we have, in this sense, a neutral satisfaction for the general organisation as well as a good satisfaction in the teaching area.

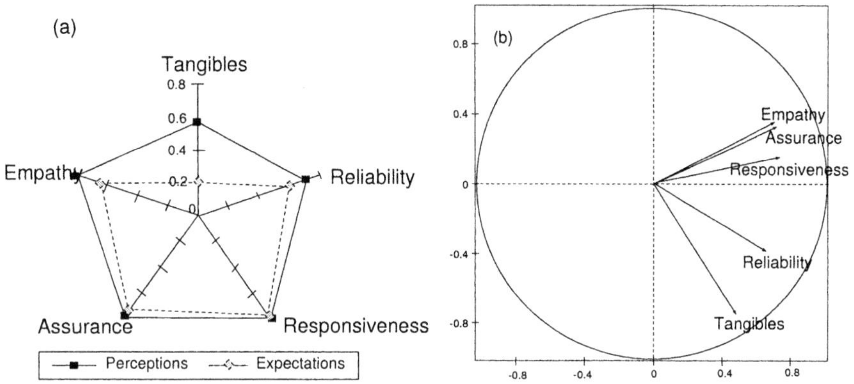

Fig. 2. (a) Radar graphic of the dimensions. (b) Interstructure phase of Statis performed to analyse the Multi-Modules version of Servperf.

The Figure 2(a) represents the radar graphic of the coordinates of the SERVQUAL dimensions on the first axis of STATICOS performed on the matrices X and Y unfolded in raw-linked variables of each dimension. We highlight as the global perceptions of each dimension are equal or greater than expectations. For the same analysis we represent the first factorial plane of the perceptions (Fig.2(b)). This figure is equivalent to the representation of the interstructure phase of Statis performed to analyse the Multi-Modules version of Servperf with unitary weights for the dimensions.

Developing the projections of the expectations and the perceptions, for each module, for statistical units on the first component of STATICOS, we obtain several diagrams (for example, here only the first 19 students), where we can read each "Multi-Modules" expectation and perception (Figure 3).We remark, in Figure 3(a) (upper left), the increasing evaluations, about the expectations, for the students 18, 5, 6 and 12 as well as the decreasing evaluations for the student 1, 9, and 15. Regarding the perceptions,

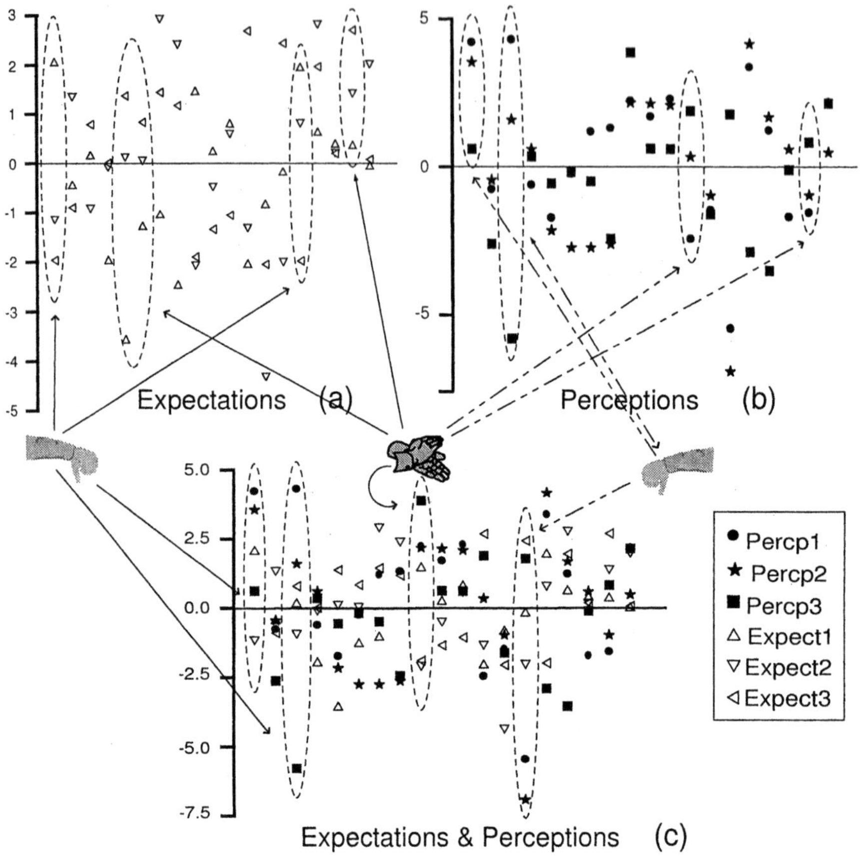

Fig. 3. Projections of the statistical units on the first axis of STATICOS.

Figure 3(b) (upper right), students 12 and 18 present increasing levels of perceptions as well as the first and the third student show the contrary. An overall representation of the multi-module expectations and perceptions can be found in figure 3(c).

In figure 4 is represented the projections of the expectations, perceptions and gaps for each module on the first factorial plane, respectively. In this case it is evident that for the expectations and perceptions (Figure 4(a) and Figure 4(b), respectively), the evaluations for the modules 1 and 2 are linked while the evaluations expressed for the module 3 is different, highlighting an overall different behaviour A similar representation can be performed for

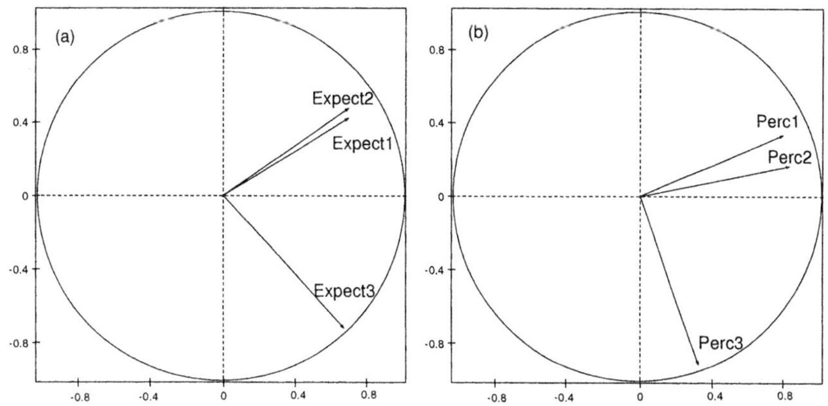

Fig. 4. (a) Expectations: 1 factorial plane. (b) Perceptions: 1 factorial plane

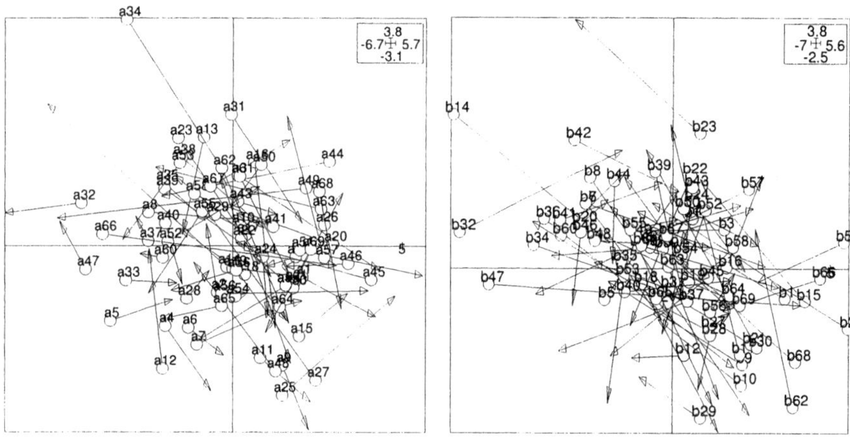

Fig. 5. Representation of the statistical units with STATICOS

the multi-module gaps. Other information about global gaps can be achieved with the representation of the statistical units with STATICOS, here for the first 2 modules, Figure 5(a) (left) and 5(b) (right) respectively. The starting point of the arrow represents the position of the statistical unit before the service has been experienced, while the end of arrow highlights the position after the service has been experienced. The length of the arrow represents the gap.

Aknowledgement: This research is financially supported by Cofin99 fund (Prof. L. D'Ambra) and with a grant Murst (ex 60%) (Prof. P. Amenta).

References

CHESSEL, D. and MERCIER, P. (1993): Couplage de triplets statistiques et liaisons especes-environement. In: J.D Lebreton and B. Asselain (eds): *Biometrie et environment*, Masson Paris.

COPPI, R. and ZANNELLA, F. (1979): L'analisi fattoriale di una serie temporale multipla relativa allo stesso insieme di unitá statistiche. In: Atti XXIX Convegno della SIS.

CRONIN J.J. and TAYLOR, S.A. (1992): Measuring service quality: a reexamination and extension, *Journal of Marketing, 56*, 55–68.

D'AMBRA, L. and AMENTA, P. (2000a): Un Approccio Statistico Multidimensionale per la Valutazione della Performance nei Servizi. In: Atti del XX Convegno Nazionale dell'Associazione Italiana Cultura Qualitá, Bologna.

D'AMBRA, L. and AMENTA, P. (2000b): Multidimensional statistical methods based on co-inertia for the 'multi-modules' customer satisfaction evaluation In: Atti del convegno CIMASI'2000 EHTP, Casablanca, Marocco.

D'AMBRA, L., AMENTA, P., GALLO M., RUBINACCI F. and SARNACCHIARO, P. (1999): Multidimensional Statistical Methods Based on Co-Inertia for the Customer Satisfaction Evaluation. In: Atti del Convegno International Quality Control, Bangkok, Thailandia.

LAFOSSE, R. (1989): Ressemblance ed difference entre deux tableaux totallement apparies *Statistique et Analyses des données*, 14, 1-24.

HANAFI, M.(1998): *Structure de l'ensemble des analyses multivariées des tableaux de données trois entres: Eléments theoriques et appliqués*, Thesé Doctoral Universite de Paris.

LAVIT, C. (1988): *Analyse conjointe de tableaux quantitatifs*. Masson Paris.

LAURO, N.C., BALBI, S., SCEPI G. (1997): L'analisi multidimensionale dei dati per la misurazione della Customer Satisfaction. In: Atti del Convegno SIS, La Statistica per le Imprese, Torino.

PARASURAMAN, A., BERRY, L. and ZEITHMAL, V. (1994): Refinement and reassessment of the Servqual scale. *Journal of Retailing, 67*, 420-450.

TORRE, F. and CHESSEL, D. (1995): Co-structure de deux tableaux totalement appariés. *Revue de Statistique Appliquée, XLIII*, 109-121.

TUCKER, L.R. (1958): An inter-battery method of factor analysis. *Psychometrika, 23*, n2, 111-136.

A Proposal of Classification of Wholesale Trade Enterprises on the Base of Structural and Performance Indicators

Paola Anitori[1] and Carlo De Gregorio[2]

[1] ISTAT, Via Balbo,16 - Roma - e-mail: anitori@istat.it
[2] ISTAT, Via Tuscolana,1060 - Roma - e-mail: cadegreg@istat.it

Abstract. In the last years wholesale trade has changed for continous new definition of wholesale trade function in the whole chain of the production-final distribution. Using a conjoint application of Principal Component Analysis and Cluster Analysis, an activity of classification of wholesale market is proposed, opposed to formal classification of Nace Rev.1, based both on the principal structure variables of concern wholesales (dimensions, local unit, location, juridical form) and on variables relative to economic accounts (invoiced, joined o added value, marginal enterprices) and export activity. Some critical points arise about the use of official classifications and also several analytical points for future studies.

1 Premise

The aim of this paper is to provide a classification of wholesale trade enterprises based on structural and performance indicators in order to highlight some peculiar aspects of the wholesale markets in Italy that the traditional classification of economic activities does not manage to stress. In particular, Principal Components Analysis (PCA) and Cluster Analysis (CA) have been performed and some different typologies among wholesale markets have been emphasised. From the economic analysis point of view, wholesale trade sector is traditionally characterised as an extremely complex one (see Consorti (1995) and Lugli (1988)), because of its role of linkage between production and retail distribution, with which it shares both synergies and growing competition. In recent years, in Italy, structural changes in retail trade and competition in manufacturing tended to vertically integrate the wholesale trade function to the point that volumes of goods traded by traditional wholesalers in some markets have drastically decreased over time. Nevertheless, even in presence of marginal enterprises, the sector didn't face any dramatic reduction in the number of operators or in the number of persons employed as it happened, for instance, in retailing. According to several authors (see Linkert (1998) and Lugli (1998)) wholesalers are transforming into services-providers and this characteristic, even if stronger in certain branches, has shown to be cross-sectional regardless to the size, localisation and specialisation of enterprises. Unfortunately, the traditional tools of analysis used also by the official statistics do not seem to give evidence of that and confine the analysis into a picture constrained by official classifications from which wholesale trade seems to stand alone in the economy and its importance is usually underestimated [1]. In this framework, for a long time, the most common description of the sector has been in terms of consistencies and comparisons with retailing with the accent on the traditional conflict between a specialised and market oriented North and a backward South of the country (see ISTAT (1998) and Zaghi (1996)). Since it seems not fully pertinent insisting on a frame in which production-oriented wholesalers of bigger size and better performance belong to northern markets whilst consumption-oriented wholesalers of small size and modest performance characterise southern markets, the approach proposed by the authors aims at finding new perspectives of analysis by means of some methodological tools left unexplored in this field.

2 Structural Aspects of Wholesale Trade in Italy

Table 1 shows the results of 1996 ISTAT Census of Industry and Services and gives a first view of the sector. Wholesalers represent one out of ten of Industry and Services firms, one out of seven of Services firms and one out of three of Distributive Trade firms. The number of persons employed is almost one million. Firms

[1] Istat has recently produced interesting insights for a critical analysis of business structure and classification. See, ISTAT (1999) ch.3 and ISTAT (2000) ch.3.

operating on a fee or contract basis represent more than 60% of total wholesale [2]. One third of the remaining firms operates in non-food consumption goods trade, while one forth operates in food and beverage trading. Intermediate goods involve one fifth of wholesale firms, while trading in investment goods regards one tenth of total enterprises. Northern regions account for more than one half of wholesalers and more than 60% of total employment, mostly concentrated in intermediate and investment goods. The average size of Italian wholesalers is 4.7 persons employed but some remarkable differences can be stressed relating to geographical and sector characteristics: central and southern wholesalers show smaller average sizes, intermediate and investment goods wholesalers are generally larger. This picture can be enriched with some economic information like turnover, value added etc. These data already show some meaningful structural characteristics of wholesale sectors. This notwithstanding, inside the three-digit Nace rev.1 breakdown strong differences among activities still remain, relating both to the specific structural characteristics assumed by upstream industry and downstream retail trade sectors structures, and to the characteristics of products in terms of storage, transportation, etc.: wholesale activities concerning fresh fruit and vegetables, meat, fish, beverages, etc., all included in the 51.3 Nace rev.1 group, are an example of this. More interesting results can be obtained through the sequential application of standard factorial methods and cluster analysis.

3 Data Analysis Applications

Using ISTAT data from structural business surveys and industry and services census, referred to year 1996, 16 structural indicators and 7 performance indicators were calculated for each of the 176 domains of study [3]. The PCA exercised on the 23 stan-

[2] The group has been excluded by the analysis proposed in this paper since the peculiar nature of their activity cannot be defined as wholesale trade in a proper sense (Zaghi (1996)). Then, from now on, the term "wholesale firms" will refer to all wholesalers but those operating in this group.

[3] The 176 domains where obtained from crossing 44 Nace Rev.1 classes - corresponding to wholesale trade activities - and 4 main geographical areas, i.e.:

Table 1. Enterprises and persons employed in wholesale trade broken down by 3-digit Nace Rev.1.

Nace Rev.1	Description	North-west	North-east	Centre	South	Total
			N. of enterprises			
51.1	Trade on a fee or contract basis	65209	53833	48913	53373	221328
51.2	Agricultural raw materials and live animals	2589	2223	1571	2510	8893
51.3	Food, beverages and tobacco	8542	6926	6594	15015	37077
51.4	Household goods	14990	9455	9386	12367	46198
51.5	Non-agricultural intermediate products, waste and scrap	10683	6900	5365	6326	29274
51.6	Machinery, equipment and supplies	5785	3831	2487	2789	14892
51.7	Other wholesale	3920	1862	1517	1853	9152
	TOTAL	111718	85030	75833	94233	366814
	% on DISTRIBUTIVE TRADE	34,9	36,7	30,1	22,3	29,9
	% on SERVICES	15,5	16,5	14,3	12,5	14,5
	% on INDUSTRY AND SERVICES	10,8	11,3	10,3	9,5	10,4
			N. of persons employed			
51.1	Trade on a fee or contract basis	85814	69407	61865	64271	281357
51.2	Agricultural raw materials and live animals	8808	7755	4400	5678	26641
51.3	Food, beverages and tobacco	45801	44791	29242	48616	168450
51.4	Household goods	88982	47507	39517	38153	214159
51.5	Non-agricultural intermediate products, waste and scrap	64234	41320	26143	23349	155046
51.6	Machinery, equipment and supplies	45849	20016	10914	7905	84684
51.7	Other wholesale	20400	7644	4710	4446	37200
	TOTAL	359888	238440	176791	192418	967537
	% on DISTRIBUTIVE TRADE	37,7	35,6	29,9	25,2	32,5
	% on SERVICES	15,5	15,1	9,2	12,5	13,2
	% on INDUSTRY AND SERVICES	7,4	7,3	5,6	7,6	7,0

Source: 1996 ISTAT Census of Industry and Services

Northeast, Northwest, Centre and South. The structural indicators where: size in terms of employed people by firm and by local unit, concentration indicators (general, per local unit, by size and by legal form), localisation indexes in terms of firms, local units and employees, average number of local units, shares of firms with multiple local units. Performance indicators where: turnover, value added, export shares on turnover, gross margin and investment per capita by two size dardised variables (60% of total inertia explained by the first two components) pointed out the weakness of bivariate correlation between enterprises' size and performance (see table 2), giving evidence of the fact that the presence of large enterprises in a domain do not imply a high performance profile in terms of profit margin or turnover per person employed. The first principal axis in fact, identified as "enterprise structure", is positively correlated to structural indicators whilst the second ("enterprise performance") is correlated with performance indicators. PCA suggested then the likelihood of the hypothesis of coexistence of different characteristics assumed by wholesale trade even within the same branches, each defined by the way structural and performance variables are combined. Figure 1, referring to Nace Rev.1 groups, gives evidence of that since the individuals (i.e.: the domains) correspond to different combination of structure and performance are split over the all graph.

Through the application of CA algorithms (this paper reports the results of the application of the Ward method performed using all the components extracted by PCA [4]), two partitions have been examined, one (explaining 50,2% of total inertia) consisting of four clusters of domains and the other - derived from the former and explaining 64% of total inertia - consisting in eight clusters mainly subsets of the main groups identified by the first partition. The four clusters partition gave us the possibility to perform a preliminary map of wholesale activities, highlighting the relations between structural characteristics, commodity specialisation and localisation. The eight clusters partition made it possible a better focusing on the previous analysis: in particular, it has been possible to establish the presence of a multitude of wholesale trade typologies even within similar markets. The first cluster gathers the domains of traditional wholesale trade, strongly characterised by small-medium sized mainly unincorporated firms, highly lo-

classes and profit ratio. For brevity, only factors with eigenvalues higher than 0.6 and variables with the absolute value of the correlation coefficients higher than 0.7 are shown in the table.

[4] The use of a limited number of the factors extracted (with eigenvalues higher than 1) has also been experimented. K-means approaches were also used. In both cases, substantially similar results were obtained.

Table 2. Description of the first two factors of PCA(1)

Factor	% Inertia	Factor 1 Variables	r	Factor 2 Variables	r
1	45.8	Enterprises. with > 20 persons employed (share)	0.93	Investment per person employed	-0.70
2	13.2	N. of persons employed	0.91	Turnover per person employed	-0.71
3	9.9	N. of persons employed in local unit	0.89	Gross operative margin/value added	-0.71
4	6.5	Persons employed in firms with > 20 persons employed (share)	0.88	Value added per pers. employed	-0.73
5	5.3	Inc. enterprises. with > 20 persons employed	0.85		
		N. of local unit per enterprise	0.76		
		Persons employed in multi-localized enterprises (share)	0.72		
		Inc. firms with ¿ 20 pers. employed (share)	0.69		
		Pers. employed in inc. firms with > 20 pers. employed (share on tot inc.firms)	0.67		

(1) For brevity, only factors with eigenvalues higher than 0.6 and variables with the absolute value of the correlation coefficients higher than 0.7 are shown in the table.

calised and with modest profit margins, trading with food and beverages (mainly fruit and vegetables), tobaccos and agriculture raw materials; a large part of southern wholesale trade employees is also represented here. Within this group, however, in the frame of the eight cluster partition, it is possible to isolate as an homogeneous subset the activities of southern wholesale trade of fresh fruit and vegetables, characterised by a strong presence of small firms, with a higher propensity to export but small profit margins.

The second cluster aggregates more organised traditional wholesale activities.

Within this group meaningful differences among domains still remain, but it is possible to split it in three subgroups. One is composed by small-medium size food wholesale activities, with modest profit margins and moderate demographic trend.

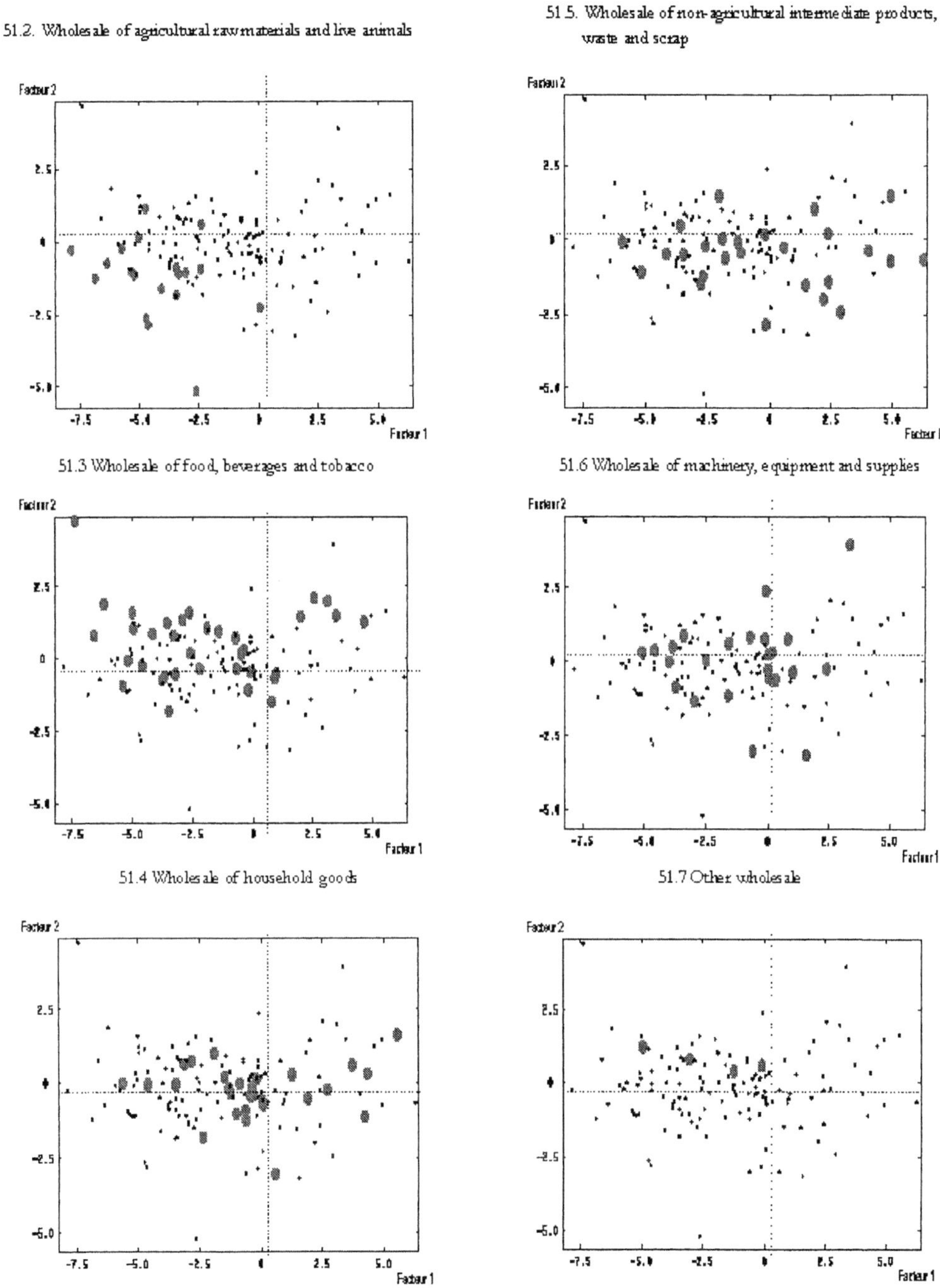

Fig. 1. Visualisation of the result of PCA

Table 3. Synoptic table of wholesale trade

3-digit Nace Rev.1	4-digit Nace Rev.1	N-W	N-E	C.	S.
Agricultural raw materials and live animals	Grain, seeds and animal feeds		2C		
	Flowers and plants	1A			
	Live animals				
	Hides, skins and leather		2B		
	Raw tobacco			2B	
Food, beverages and tobacco	Fruit and vegetables	2A		2A	1A
	Meat and meat products	2A	2A	2A	2A
	Dairy produce, eggs and edible oils and fats	2A	2A		
	Alcoholic and other beverages		2A		
	Tobacco products				
	Sugar and chocolate and sugar confectionery				
	Coffee, tea, cocoa and spices		2C		
	Non-specialised food, beverages and tobacco	4B	4A	4B	2A
	Other food, including fish, crustaceans and molluscs	2C	4B	2B	2A
Household goods	Textiles	2B	2B	2B	
	Clothing and footwear	2B	2B	2B	
	Electrical household appliances and radio and television	4A	4A	2B	2B
	China and glassware, wallpaper and cleaning materials	2C	2B	2B	
	Perfume and cosmetics	2C	2B	2B	
	Pharmaceutical goods	4A	4A	4A	
	Other household goods	2B	2B	2B	
Non-agricultural intermediate products, waste and scrap	Solid, liquid and gaseous fuels and related products		2C		2B
	Metals and metal ores	4A	4A	4A	2B
	Wood, construction materials and sanitary equipment	2B	2B	2B	
	Hardware, plumbing and heating equipment and supplies	2C	4A	2C	2B
	Chemical products	2C	2C	2C	
	Other intermediate products	2B	2B	2B	
	Waste and scrap				
Machinery, equipment and supplies	Machine-tools	2C	2B		
	Construction machinery	2C	2B	2B	
	Machinery for the textile industry	2C	2B	2B	
	Office machinery and equipment	4B	2B	2B	
	Other machinery for use in industry, trade and navigation	2C	2B	2B	
	Agricultural machinery and accessories and implements	2B	2B	2B	
Other wholesale		2B	2B		

Another subgroup is composed by the domains of the non-food wholesaler of Central and Northern regions with a medium average size, a higher propensity to export and with a more capital intensive structure. In the last subgroup firm size is larger, incorporated firms more numerous, profits margins and the other economic indicators showing better values: these are mainly northern domains, dealing with durable and intermediate goods wholesale.

The third cluster is composed by fuel wholesalers, highly specialised, with considerably high profit margins and with a stronger presence of large firms. This is a very peculiar cluster which is not possible to split further due to the extreme homogeneity of the firms both in terms of structure and performance indicators. Finally, the fourth cluster can be referred to as being composed by the most advanced wholesale activities. It can be divided in two subgroups. The first is composed by the best performing wholesale branches of the Northern regions trading with pharmaceuticals, durable goods and metals.

The second is composed by the non-specialised food wholesalers that are mainly large capital intensive firms, with a high propensity to export and lower profit margins. Table 3 summarises the results of the two partitions giving evidence of the coexistence of different typologies in each class of the Nace Rev.1. Wholesale of food, beverages and tobacco shows interesting specific characteristics explainable both on the basis of regional and sector differences. In the partitions these activities are split in the first two groups. Some domains, belonging mainly to the northern regions, are characterised by more modern business structures, and it is likely in the authors' view that this could reflect the influence of a stronger competition due to the presence of retail trade larger firms more oriented toward some sort of "self-wholesaling" activity (see Confcommercio, 1999) . In southern regions, where firm size is generally smaller, the wholesale trade of fruit and vegetables represents an extremely vital sector with a strong export activity: in northern regions, these same activities is grouped together with durable and intermediate goods wholesales activities. Meat wholesale is instead represented by very homogeneous domains in group two. The wholesale of beverages, tobacco, sugar,

coffee, etc., is characterised by a small average firm size, with the only exception of a few domains in north eastern regions.
Wholesale of household goods shows a sharp divide between southern regions and the rest of Italy. While in the south firms are generally small and poorly structured, the rest of the domains are characterised by two sets of structures: on one side, the more advanced medium firm domains, with higher profit margins and better values of the other economic indicators, on the other the small firms domains with a more traditional organisation but anyway characterised by a strong presence of incorporated firms and by good profit margins and value added per capita.
Wholesale of non-agricultural intermediate products, waste and scrap, presents strong differences with regard to business structure and performance. Both territorial and product effects influence these differences. So part of the domains (especially southern) are grouped among the more traditional and small size wholesale activities, other are grouped among the small-medium sized domains with good profit margins, and the rest present the structural characteristics of the more evolved wholesale activities.

Geographical factors explain most of the difference among wholesale of machinery, equipment and supplies. In the south small firm domains prevail. In the other regions, wholesale activities are centred on small and medium firms, mostly characterised by good profit margins. Larger firms are present in northern regions' domains. The remaining wholesale activities belong to substantially homogeneous Nace rev.1 groups.

4 Conclusions

PCA and CA revealed themselves as very useful methodological tools in analysing large economic data sets, although they are not systematically present in the economist's tool-box. In our case, they helped us to show how difficult is to base an up-to-date and coherent analysis of the sector only on the traditional approaches, mainly based on Nace rev.1 classification of economic activity. The results highlight the dramatic heterogeneity of the wholesale trade sectors, and the coexistence of different trading

Table 4. Synoptic table of wholesale trade - Legend

Clusters	Description	CODE	Characteristics
First cluster	Small size traditional wholesalers	White	Micro enterprises, low performance
Subgroup 1A	fruit and vegetables localised mainly in the South	1A	Very localised, good performance, high share of exports on turnover
Second cluster	Traditional wholesalers in consolidation		Small-medium size enterprises
Subgroup 2A	food wholesalers mainly localised in the North and Centre	2A	No local units, modest profits, low export
Subgroup 2B	medium size non-food wholesalers	2B	Good profitability, low turnover per pers.employed
Subgroup 2C	Wholesalers of durable and intermediate goods mainly localised in the North	2C	Medium-big sized, localised, medium-high profits, high turnover per pers. employed
Third cluster	Wholesalers of fuel and related products		Big sized, very high profit margins, very good performance
Fourth cluster	Advanced wholesalers		Big size, several local units
Subgroup 4A	highly advanced wholesalers in the North	4A	very high profit margins, very localised, incorporates high value added
Subgroup 4B	non-specialised food wholesalers	4B	low profit margins, low value added per pers.employed

patterns even within similar markets and products. In particular, they help to make clear the importance of the economic relations between the wholesale activities and the corresponding upstream (mainly industrial) and downstream (mainly distributive trade) sectors. Important insight have been also offered for stratification and sampling criteria to be used in business surveys dedicated to these sectors. This work represents anyway only a first step toward a deeper comprehension of the business structure of wholesale trade activities. Three further deepening seem to us extremely fruitful, one of methodological and one economic nature. On one side, in fact, significant improvements can be obtained throughout a refinement of the PCA and CA methods we applied, especially towards the recent improvements in the joint applications of these methods. secondly, it is necessary to explicitly consider the upstream and downstream business structures, linking them with the wholesale sectors' structure and building in such a way more comprehensive and general explanations. Thirdly, it seems quite promising the analysis of data for subsequent years, in order to highlight the relationship between structural changes and the results of cluster analysis.

References

BOLASCO, S., COPPI, R. (1983): *Analisi dei dati e sue applicazioni nella ricerca economica.* Edizioni Klim, Roma.

CONFCOMMERCIO (1999): *L'evoluzione della rete distributiva tra crisi e modernizzazione.* Centro studi, Servizi analisi territoriale e settoriale, Confcommercio, Roma.

CONSORTI, A. (1995): *L'economia delle aziende commerciali all'ingrosso.* Giappichelli Editore, Torino.

GAMBARO, M. (1993): Prospettive evolutive e strategie organizzative dei grossisti non alimentari italiani *Trade Marketing 4(9)*, Franco Angeli, Milano.

ISTAT (1998): *Come cambia il commercio - modificazioni strutturali e dinamica occupazionale (1980-96)*, Collana Argomenti, ISTAT, Roma.

ISTAT (1999): *Rapporto annuale 1998. La situazione del paese.* Roma.

ISTAT (2000): *Rapporto annuale 1999. La situazione del paese.* Roma.

LINKERT, K. (1998): Wholesale trade in the European Economic Area. In: *Proceedings of the Seminar on the development of commerce in Europe*, Eurostat, p.87-104.

LUGLI, G. (1988): *Efficienza del commercio e stabilit dell'industria*, Franco Angeli, Milano.

LUGLI, G. (1998) Wholesale trade in the EEA: technological innovation and transformation in the sector. In: *Proceedings of the Seminar on the development of commerce in Europe.* Eurostat, p.105-110.

MOSCUFO, M. (1999) *Indicatori di prezzo del commercio all'ingrosso: un'analisi del comparto e della sua dinamica evolutiva ai fini del calcolo di indicatori settoriali.* Serie Contributi ISTAT, n.10/99, Roma.

ZAGHI, K. (1996) La distribuzione all'ingrosso e gli intermediari del commercio. In: L. Pellegrini *La distribuzione commerciale in Italia.* Il Mulino, Bologna, p.199-233.

The Analysis of Poverty in Italy: A Fuzzy Dynamic Approach*

Daria Mendola[1] and Stefano De Cantis[2]

[1] Dipartimento di Scienze Statistiche e Matematiche, Università di Palermo, Viale delle Scienze, I-90128 Palermo, Italy
[2] Dipartimento di Metodi Quantitativi per le Scienze Umane, Università di Palermo, Viale delle Scienze, I-90128 Palermo, Italy

Abstract. The commonly used criterion to sharply separate the poor from the non poor on the basis of a poverty threshold appears to be too severe in comparison with the nature of poverty. The latter is *multidimensional* in its components (domain) and *continue* in its states (codomain). Moreover an income-based poverty line allows for a remarkable number of spurious transitions below and over that line, which do not correspond to true variations in household's standard of living. This study starts from the analysis of common (with crisp states) transition matrices; then a fuzzy multidimensional poverty indicator is built. In conclusion, fuzzy states transition matrices synthesize interpretative content of previously proposed instruments for the analysis of poverty.

1 Introduction

Poverty could seem a clear and sharp concept if we intend it as the opposite of economical richness. Nevertheless in modern approach to the measurement of poverty, this concept comprehends several aspects which involve also social dimension, among which: social marginality, poor participation in political life, dissatisfaction about one's own role in society, inadequate housing, low education, difficulties in transforming resources in "functionings" (according to Sen (1992) and (1985)). Do two persons with the same level of expenditure have the same well-being? Probably it could be not if they have different pattern of consumption: for example, we can not evaluate in the same way expenditure for medical cares and for holidays. Indeed, we could ask if two persons (having the same level of income and social opportunities) could have different capabilities in transforming economic resources in

* Although this paper is due to common work of Authors, sections 1 and 3 are attributable to S. De Cantis; sections 2, 4 and 5 to D. Mendola

well-being because of different level of education or personal ability (see Sen (1992)), or if they could have different perception of their own status. Moreover, we could ask if equidistribution of economic resources means absence of poverty or if poverty has to be contrasted because it produces social injustice or because it is a damage for economic development. Only rarely do researchers explicit theoretical assumptions underlying hypotheses and choices about relevant dimensions of concept of "poverty". We maintain, in fact, that, in analyzing this phenomenon, one would assume, implicitly or explicitly, an *economical* paradigm (to express the economical behavior of the agents and to explicit if the well-being is intended as utility, social welfare or else), a *political* paradigm (in which it must be explicated what is "equity": equidistribution, protection of weak categories, democracy, political participation, ..., and what of these finalities should be persecuted, see Hagenaars (1986)), a *sociological* theory to interpret the effect of poverty in social stratification (Abel Smith (1984)) and a *psychological* theory to detect subjective perception of one's own individual condition (role) respect of society (Hagenaars (1986) and Van Praag (1991)). Each of these paradigms has in fact a relevant impact on the choice of the components that are to be included in the concept of poverty, and on the weights we must give them. Although arbitrariness is intrinsic in applied sciences, we think that it is not a limit but a stimulus towards new specifications and developments in the research object. For more reflections upon the implications of different approaches to the study of poverty and their consequences on logical, methodological and interpretative aspects, see Mendola (2002).

2 The Approaches and the Definitions of "Poverty"

Classifying a unit (individual or household) as *poor* is a complex procedure, even if we have already chosen variables representing poverty status. In literature there are three general approaches to define poverty, known as *absolute*, *relative* and *subjective*. In the *absolute* approach poverty is intended as the condition in which

households are not able to get a minimum level of *objective* welfare (in any way we intend it) that is judged as socially acceptable. The *basic needs approach* is a way to make operative this concept. The first step consists in the selection of a set ("basic set") of goods and services that every household should posses to get an acceptable standard of living; this set is different according to the characteristics of the household members (such as age, kind of job, ...). In a second moment, we express this set by its monetary value and finally we compare this monetary threshold with household income or expenditure. If the income perceived by household is lower or equal to threshold, household is classified as "poor".
We refer to a *relative* approach when poverty is intended as an *objective* difference from some average standard of living in the society. The official approach of ISPL (*International Standard of Poverty Line*), adopted in most of industrial countries, is the most common application of the relative approach. It defines a two-members household as "poor" if its income (or expenditure) is equal or lower than the mean national *pro capite* income (expenditure). The mean national *pro capite* income (that is the "poverty threshold") is modified in order to account for household size different from two by using coefficients of an *equivalence scale*, which take account of economical scale of cohabitation. Therefore, ISPL is a unidimensional criterion based on income (or expenditure) as the unique determinant of household's welfare. This definition does not involve social deprivation or it considers the latter as included in (and well represented by) the insufficiency of income.
Recently, we noted the definition of *subjective* poverty: the perception that people have of their own situation, as an individual feeling. For lack of information about household perception on its own poverty, it is necessary to prepare suitable measurement instruments (e.g. specific questionnaires). Complexity and variety of necessary information for a subjective measure of poverty get this approach very limitedly applicable. Interesting attempts were made in Leiden Income Evaluation Project (Hagenaars and Van Praag (1985)).
The choice of a more coherent approach to measure poverty is strongly related with aim of the analysis; according to Abel Smith

(1984) we can say: "*How should poverty be defined? This question cannot be answered until we have decided for what purpose we want a definition*". In fact if poverty is cause of discontent and a damage for social cohesion, or it is perceived as a difficulty in daily managing of life, or it is only the cause of a social stratification according to poverty levels, we have to admit that only a subjective measure of poverty can take account of these aspects. Anyway, accordingly to recent literature and as far as we concern, poverty is a multidimensional phenomenon, and we find income (or expenditure) a biased indicator of poverty's status, even if we understand that in an official contest, it could be hard to interpret more complex indicators (i.e. a multidimensional ones or a subjective ones). The operationalization of the concept of poverty has hardly been agreed by researchers. Sen himself pointed out that hardly a measurement can be sharper than concept it expresses. Starting from these considerations, our study origins from the need to:

a) overcome the unidimensionality of the domain (determinants) and the dichotomy in the codomain (spaces-states) of the poverty function: the proposed indicator is in fact a synthesis of more simple indicators and it arranges deprivation state on a *continuum* according to a fuzzy approach;

b) analyze poverty in its dynamic components: this study has been carried out on a panel sample of households of which we have estimated the joint membership to fuzzy state of poverty along different periods.

Moreover, as far as concerned with our purposes, the longitudinal approach takes advantages of temporal invariance of part of the household characteristics that could act as confounding variables.

3 Poverty Analysis by Crisp States Transition Matrices

There are no surveys in Italy specifically built up to measure poverty, but only some surveys, focussed on households, that collect information about general socio-economic characteristics. The older is the Istat Annual Survey on Consumption (Istat (2000)):

it collects a detailed set of variables related to level and quality of household consumption and about main socio-demographic characteristics. The second one is the biannual Bank of Italy's Survey of Household Income and Wealth (Banca d'Italia (1997)), which put attention on the nature and the entity of household's income and patrimony. These two surveys, that are the only ones with a national relevance and a sufficient continuity during years, point out on economic characteristics of households and let out the social dimensions, necessary to complete the description of the poverty status. The latter constitutes a remarkable handicap for a correct and complete approach to the measurement of poverty. Although, since we are interested in studying the dynamic of poverty between households in a national perspective, we decided to refer our analyses to the panel component of Bank of Italy's Survey (the Istat survey doesn't have a long enough panel component). According to this, we considered only the households that participated to each survey from 1989 to 1995. It is obviously a self-selected sample but it seems to respect some representative criteria (Mendola (1999)). According to the traditional analyses of poverty, based on ISPL, we computed main classic indicators of poverty using income data and the Italian official equivalence scale (Carbonaro (1985)). Diffusion and intensity of poverty between panel households are almost stable during the first two years we considered: we found a *headcount ratio* of 6.4% and a *poverty gap ratio* (the per cent gap from poverty line of the poor) of about 20%. In the second period we note a worsening: poverty incidence rises to about 11% in 1993, and in 1995 it decreases only slightly (9.7%). Furthermore poverty is increasingly severe: the poverty gap jumps to 28.7% in 1993 and reaches 27.3% in 1995[1].

The analysis of crisp states transition matrices (that allows us to overcome, at least partially, arbitrariness just induced by the definition of the poverty line) shows that, comparing two adjacent periods, households which are stable are about 90% of the sam-

[1] The "jump" between 1991 and 1993 is perfectly consistent with the economic recession period through whom Italy went in 1992-1993. These are the years of depreciation of national money and of the exit from EMS. Also in our analyses 1993 is the worst year (about economic hardship) for Italian households.

Table 1. Permanencies and changes: transitions into and out of poverty

Transition periods		Permanence in the same state sample %	Change of state sample %	Odd Ratios (permanence vs. change)	Odds Ratios Confidence Interval ($\alpha = 0.05$)	
I lag	1989-91	93.0	7.0	21.260	11.034	40.964
	1991-93	89.6	10.4	13.876	7.616	25.282
	1993-95	90.7	9.3	22.755	13.269	39.024
II lags	1989-93	89.4	10.6	12.640	6.953	22.978
	1991-95	90.2	9.8	12.839	7.009	23.522
III lags	1989-95	88.7	11.3	7.212	3.900	13.336

ple. Consequently transition flows (see Table 1) are about 10% which is to be considered quite large: most likely some of them are spurious. But, even if the permanence and change percentages seem quite stable when the observed lag increases, the odds ratios (OR) clearly decrease. In fact we can observe that between '89 and '91 the probability to stay permanently *poor* or *not-poor* is about 21 times that to pass through a transitory hardship; when lag increases, e.g. '89-'95, the risk to change state is, as expected, substantially smaller (OR equals to 7.2). Moreover the longer the permanence in a state, the smaller the probability of change.

4 A Fuzzy Multidimensional Poverty Indicator

Building up a multidimensional indicator of poverty is a strongly arbitrary process, in which the researcher has to do sharp assumptions, principally about what are the dimensions of the phenomenon, what are their measures and also about methodologies to handle and to aggregate data information. Choice of relevant dimensions of poverty can not be argued on formal bases, this is, in fact, a researcher's option that is not confutable or falsifiable. The aggregating process is instead, without doubt, one of the statistically most relevant step in arranging a synthetic measure of poverty, and it should respect some basic statistical coherence properties. Main methodological choices, after we selected the elementary components of our indicator, are well synthesized in these steps:

1) the transformations to be operated on data to obtain an unit of measure which is common to heterogeneous components;
2) the choice of an aggregating function able to arrange elementary indicators (i.e. by sum or product) which is coherent with nature of relationship between the whole concept of poverty and its "components";
3) the selection of a sustainable weighting system according to some definite criterion. This process is not free from a high arbitrariness.

Moreover, separation between the *poor* and the *non poor* by a household-size adjusted poverty line (like ISPL), which is based on a unidimensional economic criterion, leaves us unsatisfied because it induces a remarkable number of spurious oscillations on the poverty line, due to minimum income variations which do not correspond to real changes in household welfare, especially if we refer to changes in current expenditure level. According to the multidimensional nature of poverty, we decided to decompose it into four main dimensions, constrained by available information in Bank of Italy's survey. Remarkable symptoms of the presence of poverty, that we chose as representative of the dimensions of poverty concept, are reported in Table 2 and refer to: financial hardship, housing inadequacy and subjective perception of poverty. The subjective dimension of poverty was measured according to the yes-no question: "*In 1995 did you or another member of your household consider the possibility of applying to a bank or a financial company for a loan or a mortgage but then change his/her mind thinking that the application would be rejected?*". The latter was, in our opinion, the best indicator of self-perceived poverty, available from the Bank of Italy survey.
Finally, *food-ratio* (the ratio between the expenditure for foodstuffs and the total expenditure of a household), which strictly expresses food needs but whose large interpretative potential in poverty analysis is well known, was added as a fourth dimension. Ten selected simple indicators could be arranged in many ways, in

particular using multivariate statistical methods such as Principal Component Analysis or Multidimensional Scaling, but we chose to built up a simple (linear) synthetic indicator while experimenting the interpretative potentialities of *fuzzy set theory* (Zadeh (1965), Dubois and Prade (1980)) to express simple indicators. In fact we consider fuzzy theory particularly consistent with intrinsically *continue* nature of poverty.

Let P be the population we are referring to, and let h be a house-

Table 2. The dimensions of the concept of poverty

Dimensions of poverty	**Indicators**	**Measurement levels**
financial hardship	- debts for the purchase of furniture and house's goods	yes/no
	- debts for the purchase of non-durable goods	yes/no
	- debts with friend and relatives	yes/no
housing inadequacy	- surface of house in m^2	7 ord. categ.
	- property status	3 ord. categ.
	- kind of house	6 ord. categ.
	- bathroom	yes/no
	- house heating	yes/no
subjective perception	- self-perceived poverty	yes/no
food needs	- food-ratio	(percentage)

hold in P; let A be the fuzzy subset of the poor belonging to P. A is defined by the couple: $A=\{h, f_A(h)\}$ where $f_A(\cdot)$ is called membership function of households in *fuzzy set* of the poor A. The $f_A(\cdot)$ is a function that maps the set P in the closed-limited real interval [0,1]. In particular:

$f_A(h) = 1$ means that $h \in A$ completely
$f_A(h) = 0$ means that $h \notin A$ surely.

The values of the function so that $0 < f_A(h) < 1$ mean the grades of membership of h in fuzzy set A; the higher the value of f_A for household h, the sharper its membership in fuzzy poor set. Between the membership functions available in literature, we selected the one proposed in Cheli and Lemmi (1995). So we aggregated the indicators in a composite index known as TFR (*Totally Fuzzy and Relative*) which, for household i is:

$$f(x_{i\cdot}) = \sum_{j=1}^{k} g(x_{ij})w_j \Big/ \sum_{j=1}^{k} w_j \qquad \text{i=1, 2, ..., h, ..., n} \quad (1)$$

where $g(x_{ij})$ is the grade of membership (*gom*) in poverty fuzzy set according to the j-th indicator X_j (symptom of poverty) for household i and w_j are weights which decrease as symptom diffusion increases in the population. So, $f(x_{i\cdot})$ is a weighted mean of k *goms* in fuzzy poor set.
In particular, according to TFR method:

a) $g(x_{ij})$ are based on the empiric distribution functions $H(\cdot)$ of every simple ex-post poverty indicator X_j:

$$g(x_{ij}) = \begin{cases} 0 & \text{if} \quad x_{ij} = x_j^{(1)} \\ g(x_j^{(k-1)}) + \frac{H(x_j^{(k)}) - H(x_j^{(k-1)})}{1 - H(x_j^{(1)})} & \text{if} \quad x_{ij} = x_j^{(k)}, (k > 1) \end{cases} \quad (2)$$

$x_j^{(1)}, ..., x_j^{(m)}$ representing the m categories of the variable X_j arranged in increasing order with respect to the risk of poverty, so that $x_j^{(1)}$ denotes the minimum risk and $x_j^{(m)}$ the maximum risk;

b) weights are given by $w_j = ln[1/g^*(x_j)]$ where g^* is a mean, over all households, of *goms* based on j-th indicator and represents the fuzzy proportion of the poor with respect to X_j (further methodological details in Cheli (1995)).

For the sake of brevity, we expose here only the results referring to the last two years considered, i.e. 1993 and 1995 (for more details, see Mendola (1999)). Simple indicators which had larger weights in equation (1) are about housing inadequacy: lack of house heating ($w_j = 1.79$) and the property state of the house (owned, rented, usufruct,) ($w_j = 1.33$); about economic hardship:

need of a loan to purchase non durable goods (likely for current expenditures) ($w_j = 4.52$) and to buy furniture and objects for the home ($w_j = 3.50$). Finally the food-ratio and the subjective dimension contributions are appreciable ($w_j = 3.54$). Synthetic indicators for the whole sample, that were obtained as a simple mean of (1), in the two years are equal to $P_{93} = 0.0885$ and $P_{95} = 0.0832$. Note that, to allow comparisons, we use the same weight vector for the two years, i.e. the 1993 one; anyway the indicator P_{95} made with cross-sectional weights is very similar and equals 0.0897. If we consider the TFR as a generalization of the head count ratio, we can maintain that the fuzzy proportion of poor households in 1993 equals 8.85% (instead of 11% obtained with traditional method), while in 1995 we have a fuzzy proportion of 8.2% (instead of 9.7%). The reason of this pattern is that likely the use of many indicators (of a multidimensional criterion) makes the poverty indicator more robust to income fluctuations and prevents from part of the spurious transitions mentioned above.

5 Transitions into and out of the Fuzzy Set of the Poor

For household i, let $g_{ik}^{(t)}$ be the grade of membership in state k at time t, and let $g_{ik\nu}^{(t_1,t_2)}$ be its grade of *joint membership* in state k at time t_1 and simultaneously in state ν at time t_2, which is here defined, according to the minimum entropy criterion (Manton et al. (1992)), as:

$$g_{ik\nu}^{(t_1,t_2)} = min[g_{ik}^{(t_1)}, g_{i\nu}^{(t_2)}] \qquad \text{constrained to} \begin{cases} \sum_\nu g_{ik\nu}^{(t_1,t_2)} = g_{ik}^{(t_1)} \\ \sum_k g_{ik\nu}^{(t_1,t_2)} = g_{i\nu}^{(t_2)} \end{cases} \tag{3}$$

Table 3(a) shows the matrix of *mean joint membership* of households in the two fuzzy sets of the poor (k=1) and the non-poor ($\nu = 0$) at time $t_1 = 1993$ and $t_2 = 1995$.

Analysis of mean joint membership matrix allows us to distinguish permanent and transitory poverty. In fact $E[g_{11}^{(t_1,t_2)}]$ is the fuzzy proportion of households that are poor in t_1 and in t_2, so

Table 3. Mean joint membership and fuzzy transition matrices

1993	1995		
	poor (1)	non poor (0)	
poor(1)	0.06	0.03	0.09
non poor(0)	0.02	0.89	0.91
	0.08	0.92	

(a): Mean joint membership matrix $[\mathrm{E}(g_{ik\nu}^{(1,2)})]$

1993	1995		
	poor (1)	non poor (0)	
poor (1)	0.704	0.296	1
non poor (0)	0.023	0.977	1

(b): Fuzzy transition matrix $[t_{k\nu}]$

we can assume this quantity as a *permanent poverty index*; while $\mathrm{E}[g_{10}^{(t_1,t_2)}]+\mathrm{E}[g_{01}^{(t_1,t_2)}]$ is the fuzzy proportion of households that are poor during one in two years hence they go through only a *transitory poverty*. So for the sample households, we have an index of permanent poverty equal to 0.06 (the fuzzy proportion of the poor) while the fuzzy proportion of households with only one year of poverty is 0.05 (transitory poverty index); households experimenting at least one year of deprivation represent a fuzzy proportion of sample equal to 0.11.

From the mean joint membership matrix, Tab. 3(a), we can obtain the *fuzzy transition matrix*, Tab. 3(b), whose generic element, $t_{k\nu}=\mathrm{E}[g_{ik\nu}^{(t_1,t_2)}]/\mathrm{E}[g_{ik}^{(t_1)}]$, expresses households' propensity to change state. As we can see, the "probability" (or more correctly the *propensity*) to stay poor ($t_{11}=0.704$) is rather high but considerably lower than that to stay non-poor ($t_{00}=0.977$). Moreover, Table 3(b) points out a consistent upward mobility as shown by the value $t_{10}=0.296$ (propensity to move from poverty to non-poverty) noticeably higher than the probability to move along opposite directions ($t_{01}=0.023$).

References

ABEL SMITH, B. (1984): The study and definition of poverty: values and social aims. In G. Sarpellon (ed.): *Understanding poverty.* Franco Angeli, Milano.

BANCA D'ITALIA (1997): I bilanci delle famiglie nell'anno 1995. *Supplementi al Bollettino Statistico.*

CARBONARO, G. (1985): Nota sulle scale di equivalenza. In: Commissione di Indagine sulla Povertá e sull'Emarginazione: *La povertá in Italia.* Istituto Poligrafico e Zecca dello Stato.

CHELI, B. (1995): Totally Fuzzy and Relative Measures of Poverty in Dynamic Context. An Application to British Household Panel Survey, 1991–1992. *Working Paper of the ESRC Research Centre on Micro-social Change.* Paper 95-13, Colchester, University of Essex.

CHELI, B. and LEMMI, A. (1995): A Totally Fuzzy and Relative approach to the multidimensional analysis of poverty. *Economic Notes,* 1, 115–134.

DUBOIS, D. and PRADE, H. (1980): *Fuzzy sets and systems.* Academic Press, New York.

HAGENAARS, A.J.M. (1986): *The perception of poverty.* North-Holland, Amsterdam.

HAGENAARS, A.J.M. and VAN PRAAG, B.M.S. (1985): A synthesis of poverty line definitions. *The Review of Income and Wealth,* series 31, n.2, 139–154.

ISTAT (2000): *I consumi delle Famiglie.* Istat, Roma.

MANTON, K.G., WOODBURY, M.A., STALLARD, E. and CORDER, L.S. (1992): The use of grade-of-membership techniques to estimate regression relationships. *Sociological Methodology,* 10, 309–346.

MENDOLA, D. (1999): *Le dinamiche di povertá in Italia. Un'analisi dei dati panel della Banca d'Italia con metodologie sfocate.* PhD Thesis, University of Palermo.

MENDOLA, D. (2002): Approcci, metodologie e dati per le analisi di povertá. In G. Carbonaro (ed.): *STUDI SULLA POVERTA'. Problemi di misura e analisi comparative.* Franco Angeli, Milano, 11–41.

SEN, A.K. (1985): *Commodities and Capabilities.* North Holland, Amsterdam.

SEN, A.K. (1992): *Inequality Re-examined.* Clarendon Press, Oxford.

VAN PRAAG, B.M.S. (1991): Ordinal and cardinal utility: an integration of the two dimensions of welfare concept. *Journal of Econometrics.*

ZADEH L.H. (1965): Fuzzy sets. *Information and Control, 8,* 338-353

Part V

Multivariate Analysis in Applied Sciences

Combining Information from Several Experts: Selecting and Stopping Rules in Sequential Consulting

Patrizia Agati

Dipartimento di Scienze Statistiche,Universitá degli Studi di Bologna
Via Belle Arti, 41 I-40126 Bologna, Italy. agati@stat.unibo.it

Abstract. When an investigator performs a sequential consulting of 'experts' to reduce his/her own uncertainty about an unknown quantity, efficiency reasons suggest he/she should use some criteria for selecting the 'best' expert to be consulted at each stage and for deciding at which stage the information acquiring can be stopped. In this paper some selecting and stopping rules are proposed, which are founded on some synthetic measures of the (expected) additional informative value of a not-yet-consulted expert and of the informative value of the already-acquired knowledge. A case-study, by exemplifying how the algorithms work, shows the soundness of the proposed informativeness criteria.

1 Introduction

In condition of uncertainty, an investigator can acquire information regarding an unknown quantity — the probability of an event, a risk, a future observation, a physical quantity — from heterogeneous sources of knowledge (in the following, named 'experts'): for example, research centers, information systems, theoretical or empirical models, professional opinions.

In such a context, the investigator has to face two fundamental questions:

- *how to synthesize knowledge.* Numerous algorithms have been proposed, which can be distinguished into two main classes (for a critical review, see Genest and Zidek (1986); Cooke (1991)): *axiomatic procedures*, which resolve the problem in terms of weighted averages (Laplace (1812)), and *Bayesian combinations*, where — by treating each expert's judgment as experimental data to be used in updating the investigator's prior distribution — the 'synthesis' is represented by a posterior distribution on the unknown quantity (Morris (1977);

Jouini and Clemen (1996)). The most appropriate procedure is to be chosen each time by considering various factors as, for example, the context of the problem and the sort, the quality and the amount of data available for estimating relevant parameters;

- *how to choose among options.* Often the investigator prefers to consult the experts in successive stages rather than simultaneously. So, he/she avoids wasting time (and money) by consulting a too large sample of experts: at each stage, depending on the amount of information reached, he/she can choose whether to stop or to continue the process and, depending on the answers obtained from the experts already contacted, he/she can select the 'best' expert to be consulted on the subsequent stage.

The aim of this work is to propose some selecting and stopping rules which can be suitable to be used in a sequential consulting process. The substance of such rules is almost independent of the procedure chosen for combining information from the experts; not so their mathematical form. The reference, in the present work, is the Bayesian aggregation model suggested by Morris (1977), reviewed in a recursive form: although it cannot be regarded as 'the best' combining algorithm, it's undoubted that the Bayesian paradigm offers a logically coherent answer to the expert use problem.

The paper is organized as follows. Section 2, in writing Morris' aggregation algorithm in a recursive form, gives the notation for the successive sections. In Section 3, some stopping and selecting criteria are suggested. The results from an application performed on real data, together with some concluding remarks, are presented in Section 4.

2 A Recursive Algorithm for a Sequential Belief Revision

An investigator A, who is uncertain about the value of a random quantity $\theta \in \Theta \subset \Re$, expresses his initial personal state of information in a prior probability distribution $h_0(\theta)$. For reducing his subjective uncertainty, A performs a sequential consulting of (at most n) experts Q_j: at each stage k ($k = 1, 2, \ldots, K$; $K \leq n$), the selected expert $Q^*_{j;k}$ (or, more briefly, Q_k) answers by giving a personal density $g_k(\theta)$. Treating each expert's density as result of an experiment, A can revise his own beliefs via Bayes' theorem.

Assuming that (Morris (1977)):

a) each $g_k(\cdot)$ is parameterized with a location parameter m_k and a shape parameter v_k;

b) for each k, the probability which A assigns to the event $v^{(k)} = \bigcap_{i=1}^{k} v_i$ — that is, the event "the shape parameter values the experts will give are $[v_1, ..., v_i, ..., v_k]' = \mathbf{v}$" — does not depend on θ: in symbols, $\ell\left(v^{(k)}|\theta\right) =$ $= \ell\left(v^{(k)}\right)$;

Morris shows that the posterior density can be written as[1],

$$h\left(\theta|m^{(k)}, v^{(k)}\right) = \frac{\ell\left(m^{(k)}|v^{(k)}, \theta\right) \cdot h_0(\theta)}{\int_\Theta \ell\left(m^{(k)}|v^{(k)}, \theta\right) \cdot h_0(\theta)\, d\theta} \tag{1}$$

where:

- $\ell\left(m^{(k)}|v^{(k)}, \theta\right)$, denoted in the following by $\ell_k(\theta)$ for notational convenience, indicates the subjective conditioned likelihood function of θ for the data $m^{(k)} = \bigcap_{i=1}^{k} m_i$, given $v^{(k)}$: it represents — for θ varying — A's probabilities that the location parameter values the experts will provide are $\mathbf{m} = [m_i]'_{i=1,\ldots,k}$;
- the posterior $h\left(\theta|m^{(k)}, v^{(k)}\right)$ or, more briefly, $h_k(\theta)$, represents A's (subjective) synthesis distribution at stage k.

If the following assumption holds too,

c) for each k, the conditional probability which A assigns to the event "$g_k(\cdot)$ shape parameter will be v_k", given $m^{(k-1)}, v^{(k-1)}$ and θ, does not depend on θ: that is, $\ell\left(v_k|m^{(k-1)}, v^{(k-1)}, \theta\right) = \ell\left(v_k|m^{(k-1)}, v^{(k-1)}\right)$;

[1] It can be shown that these assumptions can be relaxed without changing substantially the results (Morris (1977)).

then Morris' (simultaneous) aggregation algorithm (1) can be written in a recursive form as,

$$h_k(\theta) = \frac{\ell\left(m_k | v_k, m^{(k-1)}, v^{(k-1)}, \theta\right) \cdot h_{k-1}(\theta)}{\int_\Theta \ell\left(m_k | v_k, m^{(k-1)}, v^{(k-1)}, \theta\right) \cdot h_{k-1}(\theta)\, d\theta} \tag{2}$$

where $\ell\left(m_k | v_k, m^{(k-1)}, v^{(k-1)}, \theta\right)$ is the subjective conditioned likelihood function of θ for the only observation m_k, given v_k and also the location and shape values provided by the $k-1$ previously consulted experts.

As regards the arduous assessment of the function $\ell(\cdot)$ in (2), the relation $\ell\left(m_k | v_k, m^{(k-1)}, v^{(k-1)}, \theta\right) = \ell_k(\theta) / \ell_{k-1}(\theta)$ allows to use Morris' (simultaneous) result,

$$\ell_k(\theta) \propto C_k(\theta) \cdot \prod_{i=1}^{k} g_i(\theta) \tag{3}$$

where the subjective *calibration function* $C_k(\theta)$ encapsulates the investigator's state of knowledge about each expert's probability assessment ability and the degree of dependence among the k experts.

In short (for details, see Morris (1977)), let τ_i denote the i-th *performance indicator*, defined as Q_i's cumulative function $G_i(\cdot | m_i, v_i)$ evaluated at the true value of θ: $C_k(\theta)$ expresses the admissibility degrees which A assigns to each possible θ value looked at as the realization of the k-dimensional quantile vector $\tau = [\tau_i]'_{i=1,\ldots,k}$. Technically, $C_k(\cdot)$ is nothing but a subjectively assessed density $\phi_k(\cdot)$ of τ, conditioned on $\mathbf{v}$ and θ, looked at as a function of θ (for fixed $\mathbf{m}$): in symbols, the relation between the so-called *performance function* $\phi_k(\cdot)$ and the calibration function $C_k(\theta)$ is,

$$\phi_k(\tau | \mathbf{v}, \theta) = \phi_k\left[\mathbf{G}(\theta | \mathbf{m}, \mathbf{v}) | \mathbf{v}, \theta\right] = C_k(\theta) \tag{4}$$

where $\mathbf{G}(\theta) = \mathbf{G}(\theta | \mathbf{m}, \mathbf{v})$ denotes the vector $[G_i(\theta | m_i, v_i)]'_{i=1,\ldots,k}$.

Whenever the investigator has only some pieces of information about the experts — an 'information block' which is not adequate to construct an empirically founded probability distribution of their performance indicators — the fiducial argument (Fisher, 1956) can be used for inductively modelling the calibration function, enabling it to be specified with a relatively small number of assessments and warranting A's beliefs about the experts in terms of personal long run frequencies (Monari and Agati (2001)).

With the following notation:

- $\tilde{\mathbf{G}}(\theta) = \left[\tilde{G}_i(\theta)\right]'_{i=1,\dots,k}$, with $\tilde{G}_i(\theta) = \ln\left[G_i(\theta)/(1 - G_i(\theta))\right]$;
- $\tilde{\mathbf{t}} = \left[\tilde{t}_i\right]'_{i=1,\dots,k}$, with $\tilde{t}_i = \ln\left[t_i/(1 - t_i)\right]$;
- c as normalization constant;

the resulting fiducial calibration function can be written as,

$$C_k(\theta) = C_k(\theta; \mathbf{t}, \mathbf{S}) =$$

$$= c \cdot \prod_{i=1}^{k} \{G_i(\theta) \cdot [1 - G_i(\theta)]\}^{-1} \cdot \exp\left\{-\frac{1}{2}\left[\tilde{\mathbf{G}}(\theta) - \tilde{\mathbf{t}}\right]' \mathbf{S}^{-1} \left[\tilde{\mathbf{G}}(\theta) - \tilde{\mathbf{t}}\right]\right\} \quad (5)$$

It's worth noting that the function (5) is univocally defined by:

- A's assessment $\mathbf{t} = [t_i]'_{i=1,\dots,k}$ of the performance indicator τ;
- the subjective variance-covariance matrix $\mathbf{S}$, reflecting A's information about the variability and the reciprocal dependence of the experts' performance indicators.

3 Selecting and Stopping Rules

The purpose of expert consulting is reducing the investigator's uncertainty about the unknown quantity θ. So, in designing and performing the sequential process, it is reasonable to found the selecting and stopping rules on some criterion of informativeness.

In particular, though no single number can convey the amount of information encapsulated in a density function, a synthetic measure of the (*expected*) additional informative value of a not-yet-consulted expert $Q_{j;k}$ is indispensable for selecting the one to be consulted at stage k, especially when A's calibration assessments, together with the shape parameters provided by the experts, lead to not-coinciding preference orderings. And, analogously, as likelihood functions and posterior densities can display a wide variety of form, a synthetic measure of the reached knowledge degree about θ is needed for picking out the 'optimal' stage k^* at which data acquiring can be stopped.

Suppose the investigator A is performing the process of revising beliefs in light of new data according to the algorithm described in Section 1. He has specified his prior $h_0(\theta)$; each of n contacted experts Q_j has revealed the variance v_j — assumed as uninformative about θ: see *b)* in Section 2) — of his own density $g_j(\theta)$, and A has already consulted $k-1$ of them, so obtaining the locations of $k-1$ expert densities: he is now at stage k of the process (Figure 1), and must select one among the not-yet-consulted experts $Q_{j;k}$ $(j = 1, 2, \ldots, n-k+1)$.

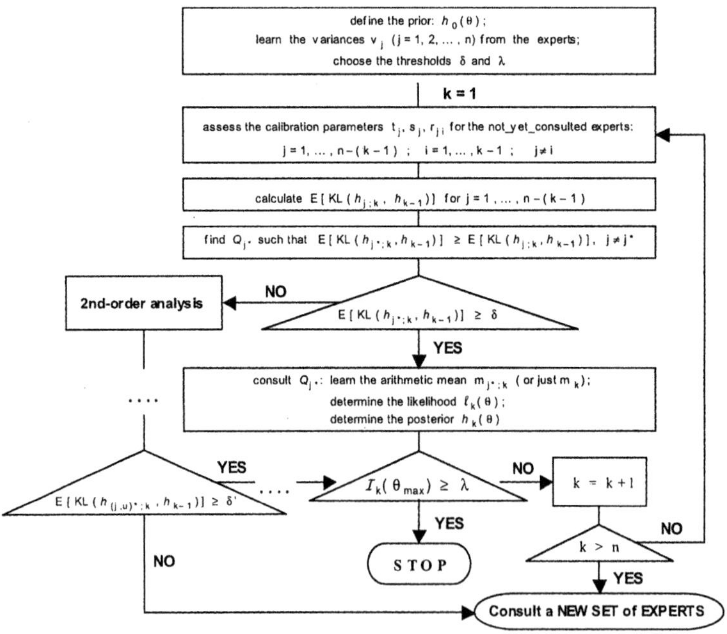

Fig. 1. Flow-chart of the sequential procedure.

For each $Q_{j;k}$, the investigator A assesses — conditionally on v_j, on the basis of the information at his disposal (including all the expert locations m_i revealed up to stage $k-1$) — the parameters of the k-stage calibration function $C_{j;k}(\theta)$: that is, t_j, s_{jj} and the covariances s_{ji} (or the linear correlations r_{ji}) between $Q_{j;k}$ and each already-consulted expert Q_i, $i = 1, 2, \ldots, k-1$.

At this point of the procedure, no $Q_{j;k}$ has revealed the location value m_j of his own $g_j(\theta)$: the several 'answers' m_j which each can virtually give are not all equally informative, so the (informative) value of each expert at the k-stage — to be measured with regard to A's current knowledge[2] of θ reflected in the posterior density $h_{k-1}(\theta)$ of the previous stage — is an *expected* value, calculated by averaging a selected measure of relevant information about θ in $Q_{j;k}$'s answer over the space M_j of the virtually possible m_j values.

By reasoning in a *knowledge* context — which is an *inductive* context, where an expert opinion is more relevant the more it is able to modify the investigator's posterior distribution on the unknown quantity — a suitable measure of $Q_{j;k}$'s informative value can be the *expected Kullback-Leibler divergence* of the density $h_{j;k}(\theta)$ with respect to the previous stage posterior $h_{k-1}(\theta)$,

$$\mathrm{E}\left[KL\left(h_{j;k}, h_{k-1}\right)\right] := \int_{M_j} f\left(m_{j;k} \middle| v_{j;k}, m^{(k-1)}, v^{(k-1)}\right) \cdot KL\left(h_{j;k}, h_{k-1}\right) dm_j \tag{6}$$

where the KL-divergence (Kullback (1959)),

$$KL\left(h_{j;k}, h_{k-1}\right) := \int_{\Theta} h_{j;k}(\theta) \cdot \ln\left[h_{j;k}(\theta) / h_{k-1}(\theta)\right] d\theta \tag{7}$$

measures indirectly the information provided by an answer $m_{j;k}$ in terms of the changes it yields on the density $h_{k-1}(\theta)$. The conditional density $f(\cdot)$ in (6) is equal to the denominator of (2) read as a function of $m_{j;k}$ and normalized; when assumptions *a)*,

---

[2] In fact, all the other elements being equal, the more A is uncertain about θ, the more an answer m_j is worthy.

b) and *c)* hold, it can be determined as

$$f\left(m_{j;k}|v_{j;k}, m^{(k-1)}, v^{(k-1)}\right) = f\left(m^{(j;k)}|, v^{(j;k)}\right) / f\left(m^{(k-1)}|v^{(k-1)}\right) \tag{8}$$

where the density $f\left(m^{(j;k)}|, v^{(j;k)}\right)$ — and analogously $f\left(m^{(k-1)} \mid v^{(k-1)}\right)$ — is equal, up to the normalization term, to the denominator $\int_{\Theta} \ell\left(m^{(k)}|v^{(k)}, \theta\right) \cdot h_0(\theta)\, d\theta$ of (1), read as a function of $m^{(k)}$.

The expert $Q^*_{j;k}$ presenting the greatest expected KL divergence is, at stage k, the most informative: but is he an expert worth consulting? The answer is yes, if the information he provides is, on average, *enough* different from what A already knows about θ, *i.e.* if the expected divergence of $h_{j^*;k}(\theta)$ with respect to $h_{k-1}(\theta)$ is not less than a predetermined value δ $(0 \leq \delta < \infty)$. About the choose of the threshold δ, a very useful tool is the 'calibration scheme' proposed by McCulloch for deciding whether a KL-divergence value is a large or a small one (see McCulloch (1989)).

So the *selecting rule* can be expressed as follows. *Consult the expert* $Q^*_{j;k}$ *such that*

$$\mathrm{E}\left[KL\left(h_{j^*;k}, h_{k-1}\right)\right] \geq \mathrm{E}\left[KL\left(h_{j;k}, h_{k-1}\right)\right] \qquad j \neq j^* \tag{9}$$

on condition that

$$\mathrm{E}\left[KL\left(h_{j^*;k}, h_{k-1}\right)\right] \geq \delta \tag{10}$$

If $Q^*_{j;k}$ *does not satisfy* (10), *then proceed to a 2-th order analysis: that is, consult the pair* $\left(Q_{j;k}, Q_{u;k}\right)^*$ *presenting the greatest expected KL-divergence, provided that it is* $\mathrm{E}\left[KL\left(h_{(j,u)^*;k}, h_{k-1}\right)\right] \geq \delta$*; otherwise contact a new set of experts and perform a new process by using the posterior* $h_{k-1}(\theta)$ *as a new prior* $h_0^{'}(\theta)$.

The expert $Q^*_{j;k}$ satisfying (10) becomes just Q_k, the "k-stage expert". By consulting him, A learns the location m_k of the density $g_k(\cdot)$: now, the k-stage calibration function $C_k(\theta)$ is univocally defined, and consequently, the likelihood function $\ell_k(\theta)$ and the posterior density $h_k(\theta)$ too.

In theory, the investigator should stop the consulting only when the aggregated knowledge about θ, reflected in the posterior density, is 'inertially stable': *i.e.*, only when additional experts, even if jointly considered, are not able to modify appreciably the synthesis distribution, on the contrary they contribute to its inertness. But too many experts could be needed for realizing such a stopping condition. It can be weakened by requiring just the knowledge about θ deriving from expert answers to be enough for A's purposes. A measure encapsulating the strength of the experimental data in determining a preference ordering among 'infinitesimally close' values of θ is Fisher's notion of information. The value of the *observed information* $I(\cdot)$ at the maximum of the log-likelihood function,

$$I_k(\theta_{\max}) := -\partial^2/\partial\theta^2 \ln \ell_k(\theta_{\max}) \tag{11}$$

is a second-order estimate of the spherical curvature of the function at its maximum: within a second-order approximation, it corresponds to the KL-divergence between two distributions that belong to the same parametric family and differ infinitesimally over the parameter space.

So, the *stopping rule* may be defined as follows. *Stop the consulting at stage k^* at which a pre-selected observed curvature λ of the log-likelihood*[3] *valued at $\theta := \theta_{\max}$ has been reached,*

$$I_{k^*}(\theta_{\max}) \geq \lambda \tag{12}$$

For deciding whether a curvature value $I(\theta_{\max}) = w$ is a large or a small one, a 'calibration' can be performed by thinking of a binomial experiment where a number $x = n/2$ of successes is observed in n trials and finding x such that $I(\hat{p}_{ML} = 0.5) = w$, where $\hat{p}_{ML} = 0.5$ is the maximum likelihood estimate of the binomial parameter p. Table 1 shows a range of x values with the corresponding w curvature values. The simple relation $x = w/8$ holds: so, for example, if $w = 120$, the width of the curve $\ln \ell_k(\theta)$ near $\theta := \theta_{\max}$ is the same as the curve $\ln \ell(p)$ at $\hat{p}_{ML} = 0.5$ when $x = 15$ and $n = 30$.

[3] It can be used also the curvature of log-posterior, so that the whole knowledge, including the prior one, is considered.

Table 1. 'Calibration' of curvature values: relation between x and w values.

x	1	2	5	10	15	20	25	30	40	50
w	8	16	40	80	120	160	200	240	320	400

4 Case-study and Concluding Remarks

The behavior of the algorithms proposed in the previous section — and implemented (Agati and Stracqualursi (2001)) in MATHEMATICA — has been investigated in simulation and experimental studies.

In this section, the results from medical data are synthetically[4] presented to exemplify how the selecting and stopping rules work. Particularly, data in Table 2 regard a sequential consulting process of $n = 4$ orthopaedists, performed by an Italian research laboratory about the long-term failure log-odds θ of a new hip prosthesis. A fifth surgeon — which play the role of investigator — has assessed the calibration parameters, without modifying them in proceeding from a stage to the successive one. He has also (subjectively) chosen the following thresholds:

- $\delta = 0.02$: by reading this value in McCulloch's scale, at stage k the most informative expert $Q^*_{j;k}$ is consulted only if the expected KL-divergence of $h_{j^*;k}(\theta)$ with respect to $h_{k-1}(\theta)$ is not less than the KL-divergence of a Bernoulli distribution $B(p)$ with $p = 0{,}5$ from a Bernoulli distribution with $p = 0.65$; or, in other words, only if stopping the process at stage $k-1$ instead of proceeding to stage k involves, on average, an information loss larger than that one yielded by using a $B(0{,}65)$ instead of a $B(0{,}5)$;
- $\lambda = 120$: by using the scale proposed in Section 3, the consulting process is stopped at stage k^* at which the observed curvature of the log-likelihood function $\ln \ell(\theta)$ valued at $\theta := \theta_{\max}$ is the same as the function $\ln \ell(p)$ at $\hat{p}_{ML} = 0.5$ when, in a binomial experiment, $n = 30$ and $x = 15$.

[4] A more detailed discussion about relevant methodological and technical issues involved in applications concerning data of this kind is the subject of a paper in preparation.

Table 2. Input data for the sequential consulting of four orthopaedists about long-term failure log-odds of a new hip prosthesis.

Q_j	v_j	t_j	s_{jj}	r_{j1}	r_{j2}	r_{j3}	r_{j4}
Q_1	0.150	0.45	1.20	1			
Q_2	0.145	0.65	1.50	+0.20	1		
Q_3	0.120	0.75	1.70	−0.05	+0.50	1	
Q_4	0.110	0.45	1.10	+0.10	+0.10	+0.10	1

In this study, the conditions *a)*, *b)* and *c)* mentioned in Section 2 can be held to be satisfied. In fact: *a)* it rests on empirical evidence — and the experts confirm it — that the failure log-odds θ can be supposed as Gaussian; *b)* it is reasonable to think the probability the fifth orthopaedist assigns to the event "the experts will give the variances $[v_1, ..., v_4]' = \mathbf{v}$" is the same for all θ values: so the surgeons' stated variances alone give no information able to change the investigator's beliefs about θ; *c)* it is reasonable as well to assume the conditional probability the investigator assigns to the event "the expert $Q_{j;k}$ will give the variance v_k", given the shape and location values provided by the $k-1$ previously consulted experts, is the same for all θ values. So the combining algorithm outlined in Section 2 has been applied, as well as the selecting and stopping rules suggested in Section 3.

Table 3. Output of the proposed sequential procedure in the consulting of four orthopaedists about long-term failure log-odds of a new hip prosthesis.

Q_j	*Stage* $k=1$ $\mathrm{E}[KL(h_{j;1}, h_0)]$	*Stage* $k=2$ $\mathrm{E}[KL(h_{j;2}, h_1)]$	*Stage* $k=3$ $\mathrm{E}[KL(h_{j;3}, h_2)]$
Q_1	1.41487	**1.92935**	—
Q_2	1.35582	1.52293	1.42842
Q_3	1.42427	1.72981	**1.93624**
Q_4	**1.60348**	—	—
	↓	↓	↓
$Q^*_{j;k}$	Q_4	Q_1	Q_3
m_k	−1.208	−1.992	−2.752
$I_k(\theta_{\max})$	18.713	53.492	138.984 ($> 120 = \lambda$)

Table 3 summarizes the results of the sequential process, while Figure 2 shows the posterior distributions $h_k(\theta)$ at each stage.

For $k = 1$, the selecting rule proposed in Section 2 chooses the expert Q_4: really he offers the smallest variance ($v_4 = 0.110$), and also the investigator's uncertainty about his performance indicator is assessed to be the smallest ($s_{44} = 1.10$). Q_4's answer ($m_{4;1} = -1,208$) leads to a curvature value $I_1(\theta_{\max}) = 18.713 < 120 = \lambda$: so the process goes on. At stage 2, the selecting rule shows its usefulness: in fact, the v_j, t_j and s_{jj} values[5] don't lead to a unique preference ordering. The most informative expert Q_1 is selected by the algorithm[6] and $m_{1;2}$ is observed. The curvature value is $I_2(\theta_{\max}) = 53.492 < 120 = \lambda$: the consulting proceeds. At stage 3,

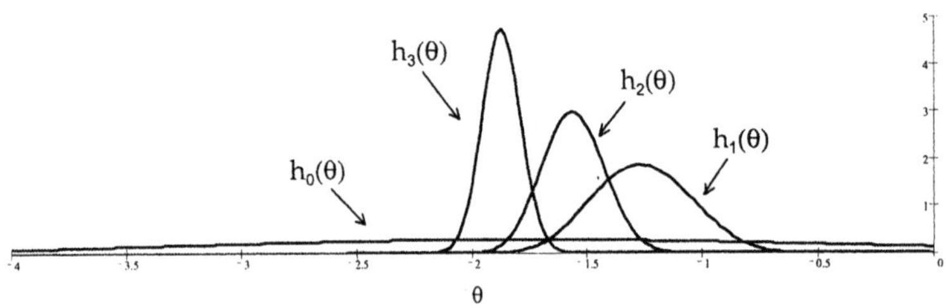

Fig. 2. Investigator's distributions at stages 0 (*i.e.*, the prior), 1, 2 and 3 of the sequential procedure.

the preference for Q_3 instead of Q_2 is also (but not only) motivated by the correlations with Q_1: a negative correlation ($r_{31} = -0.05$) is more informative than a weak positive one ($r_{21} = 0.20$). The observed $m_{3,3}$ leads to $I_3(\theta_{\max}) = 138.984 > 120 = \lambda$. The process is stopped: the expert Q_2 is left out of the consulting and the stage-3rd posterior $h_3(\theta)$ — whose location and shape values are, respectively, -1.873 (the median, here coinciding with the arithmetic mean and the mode) and 0.084 (the standard deviation) — can be regarded as the synthesis expression of the expert

[5] The correlations between Q_4 and the other experts are all equals: so they don't come into play.

[6] It's worth noting that the value $m_{4;1}$ observed at stage 1 has modified, at stage 2, the previous-stage preference ordering: for this reason, the selecting at each stage one only expert is to be preferred to selecting a set of experts (simultaneously).

knowledge about the long-term failure log-odds θ of the new hip prostheses.

By looking at the selecting and stopping output of the case-study, the behavior of the informativeness criteria appears to be coherent with the intuition, so giving an empirical support about the soundness of the proposed selecting and stopping algorithms in performing an efficient sequential consulting process.

References

AGATI, P. (2001): *La sintesi di informazioni esperte. Aspetti metodologici e applicativi in ambito decisionale e scientifico.* Clueb, Bologna.

AGATI, P. and STRACQUALURSI, L. (2001): *Algoritmi computazionali per l'aggregazione di opinioni esperte.* Quaderno del Dip. di Scienze Statistiche "Paolo Fortunati", Università di Bologna, Serie Ricerche n. 4, Clueb, Bologna.

COOKE, R. M. (1991): *Experts in Uncertainty. Opinion and subjective probability in science.* Oxford University Press, Oxford.

FISHER, R. A. (1956, second edition 1959): *Statistical Methods and Scientific Inference.* Oliver & Boyd, Edinburgh.

GENEST, C. and ZIDEK, J. V. (1986): Combining Probability Distributions: a Critique and an Annotated Bibliography. *Statistical Science, 1*, 114–148.

JOUINI, M. N. and CLEMEN, R. T. (1996): Copula Models for Aggregating Expert Opinions. *Operations Research, 44*, 444–457.

LAPLACE, P. S. (1812): Théorie analytique des probabilities. In: *Oeuvres completes de Laplace*, vol. VII, Gauthier-Villars, Paris.

McCULLOCH, R. E. (1989): Local Model Influence. *Journal of the American Statistical Association, 84*, 473–478.

KULLBACK, S. (1959): *Information Theory and Statistics.* Wiley, New York.

MONARI, P. and AGATI, P. (2001): Fiducial Inference in Combining Expert Judgements. *Statistical Methods and Applications. Journal of the Italian Statistical Society, 10*, 81–97.

MORRIS, P. A. (1977): Combining Expert Judgments: a Bayesian Approach. *Management Science, 23*, 679–693.

A Spatial Clustering Hierarchical Model for Disease Mapping

Massimo Bilancia and Alessio Pollice

Dipartimento di Scienze Statistiche, Università degli Studi di Bari - Via Camillo Rosalba n. 53, 70124, Bari, Italy.
E-mail: mabil@dss.uniba.it, Tel. +39/(0)805049341, Fax +39/(0)805049147

Abstract. In this paper a finite mixture model with a specific weights for each observation is introduced. The logistic transformation of these weights is modelled through a markovian field, with space autocorrelations of Gaussian type. This specification is particularly useful for desease mapping issues: some implementation difficulties are shortly discussed, together with the problem of the choice of the mixture's components number.

1 Introduction

Disease mapping is the representation and analysis of maps of disease incidence or mortality data. This issue has been recently addressed in the framework of Bayesian hierarchical modelling, considering a Poisson model at the lowest level of the hierarchy

$$[y_i|\theta] \sim \text{Poisson}\,(\theta_i E_i) \tag{1}$$

where y_i is the count of new cases or deaths in a given period for a given pathology in the ith of n areas, E_i is the expected count we would observe under a standard rate, and θ_i is the unobserved (true) relative risk; spatial heterogeneity usually occurs when count data are referred to small areas and/or rare diseases.

The common formulation of the higher level of the hierarchy (Besag et al. (1991), Breslow and Clayton (1993)) borrows its structure from the framework of the Generalized Linear Mixed Model (GLMM), being defined by

$$\varphi_i = u_i + \varepsilon_i, \quad i = 1, \ldots, n \tag{2}$$

where the log-relative risks $\varphi_i = \log \theta_i$ are modelled by means of two mutually independent random effects: ε_is that capture unstructured (extra-Poissonian) variation, being distributed as

$$[\varepsilon_i|\gamma_\varepsilon] \overset{i.i.d.}{\sim} \mathrm{N}\left(0,\gamma_\varepsilon^{-1}\right), \quad i=1,\ldots,n \tag{3}$$

and u_is that model the heterogeneity caused by unmeasured covariates (risk factors) inducing spatial correlation among counts of nearby areas. A popular joint specification relies on the Conditional Autoregressive (CAR) prior for $u=(u_1,\ldots,u_n)$ (Besag and Kooperberg, 1995), given by the following improper density

$$[u|\gamma_u] \propto (\gamma_u)^{n/2}\exp\left[-\frac{\gamma_u}{2}\sum_i\sum_{s\neq i} w_{is}\left(u_i-u_s\right)^2\right] \tag{4}$$

where weights w_{is} are generally specified according to the *adiacency model* with $w_{is}=1$ if area i and s share a common boundary and 0 otherwise. Bayesian model specification is completed assigning a suitable prior distribution to parameters γ_ε and γ_u: see, for example, Bilancia and Pollice (2000) for details.

Many authors noticed that the above mentioned BYM model (Besag, York and Molliè) may induce oversmoothing when the underlying marginal distribution of the relative risk is discontinuous, including a finite number of well separated levels of risk. To overcome this issue, Knorr-Held and Raßer (2000) introduce a model providing a definite segmentation of the space of relative risks based on a modification of Voronoi tessellations; in a similar fashion, Green and Richardson (2000) describe the heterogeneity of observed counts by means of a finite mixture models with spatially autocorrelated allocation counts modelled by a Potts process. A companion paper of the latter is due to Fernandez and Green (2000), where alternative allocation mechanisms are explored.

All the models described in the aforementioned papers are estimated by the Reversible Jump MCMC technology, and thus are hardly amenable to the everyday use by researchers not experienced in the subject. In this paper we propose a simpler model based on the hierarchical specification described in Pollice and Bilancia (2000), which can be easily implemented using the multipurpose package for Bayesian analysis Winbugs 1.3 (Spiegelhalter et al. (2000)).

2 The Model

Following Pollice and Bilancia (2000), the possibly discontinuous risk structure (k levels of risk) is taken into account by assuming that counts are conditionally independent among areas and distributed according to the following k-components mixture model having area-specific component weights

$$[y_i|\psi, z] \overset{i.i.d.}{\sim} \text{Poisson}\left(\psi_{z_i} E_i\right), \quad i = 1, \ldots, n \tag{5}$$

$$\Pr\left(z_i = j|\pi\right) = \pi_{ij}, \quad j = 1, \ldots, k, \quad \text{ind. over } i = 1, \ldots, n \tag{6}$$

with the obvious restriction $\sum_{j=1}^{k} \pi_{ij} = 1$. The dependence of mixture weights on the explanatory variables is taken into account by the following multicategorical logit model considering the kth category as a reference category

$$\log\left(\pi_{ij}/\pi_{ik}\right) = \eta_{ij}, \quad i = 1, \ldots, n, \quad j = 1, \ldots, k \tag{7}$$

Expression (7) can be reformulated as

$$\pi_{ij} = \frac{I_{[j=k]} + \left(1 - I_{[j=k]}\right) \exp\left(\eta_{ij}\right)}{1 + \sum_{h=1}^{k-1} \exp\left(\eta_{ih}\right)} \tag{8}$$

where $I_{[j=k]}$ is the indicator function of the kth category (for which the linear predictor is identically null). Spatial dependence is achieved modelling the weights π_{ij} via the linear predictor η_{ij} taking the following form

$$\eta_{ij} = \beta_0 + u_{ij}, \quad i = 1, \ldots, n, \quad j = 1, \ldots, k-1 \tag{9}$$

where $u_j = (u_{1j}, \ldots, u_{nj})$ is given a first order CAR prior (4)

$$[u_j|\gamma_u] \propto (\gamma_u)^{n/2} \exp\left[-\frac{\gamma_u}{2} \sum_i \sum_{s \in \partial i} \left(u_{ij} - u_{sj}\right)^2\right] \tag{10}$$

In the above formula ∂i denotes the set of first order neighbours of area i. Model specification is completed by assigning a flat prior to the common intercept β_0, i.e.

$$[\beta_0] \propto 1 \tag{11}$$

Note that the parameter β_0 has to be included in the model when using Winbugs 1.3 for its estimation: the flat specification (11) is compulsory as well. When such a constant term is included in the linear predictor, an identifiability sum-to-zero constraint $\sum_{i=1}^{n} u_{ij} = 0$ must be introduced: Winbugs 1.3 automatically imposes it " on the fly" by recentering the $u_{ij,(g)}$ samples around their own mean at every MCMC iteration g.

Finally, a usual specification for the prior distribution of the smoothing parameter γ_u is given by

$$[\gamma_u] \sim \text{Gamma}\,(a_u, b_u) \tag{12}$$

Hyperparameters a_u and b_u are assumed to be known: their setting is notoriously difficult and some sensitivity analysis is generally required. A similar specification is used for the inverse variance γ_ε of unstructured random effects ε_i in the BYM model.

According to the Bayesian clustering framework, the ith area is assigned to the $j^\star$th level of risk when the posterior distribution of the allocation variable z_i reaches its mode at $j^\star$, i.e. when

$$\text{argmax}_j \text{Pr}\,(z_i = j|y) = j^\star \tag{13}$$

After performing the classification of all areas on the basis of a MCMC output-based approximations of (13), maps can be drawn by coloring individual areas accordingly: the resulting choroplet is usually very informative for public health assessment and monitoring.

Other meaningful posterior summaries that can be easily estimated by the MCMC output are the posterior mean of the jth level of risk $\text{E}(\psi_j|y)$, and the posterior expected relative relative risk of area i $\text{E}(\psi_{z_i}|y)$ (notice that the proposed hierarchical specification does not contain area-specific relative risks in the likelihood).

A suitable choice of the prior distribution of relative risks is

$$[\psi_j] \overset{ind.\,j}{\sim} \text{Gamma}\,(\alpha, \beta) \tag{14}$$

The choice of hyperparameters in (14) is not an intuitive task and deserves some attention: a central difficulty with Poissonian mixtures stems from the fact that it is not possible to be informative on the prior distribution of ψ_j, for the obvious reason that ψ_j is both the mean and the variance of the distribution of the counts. Several authors discuss weakly informative settings: among these Dellaportas et al. (1997) choose $\alpha = \beta = 0.01$. However, the latter prior is sharply peaked in a neighborhood of zero, favoring the inclusion of hardly identifiable components in the posterior distribution (furthermore, for the dataset presented in this paper the use of a large prior variance led to unstable and switching MCMC simulations). We rather adopt a default option introduced in Viallefont et al. (2000), by setting $\alpha > 1$ to avoid the aforementioned peak effect, and the prior mean α/β equal to the median of observed Standardized Mortality Ratios.

3 Model Choice

To compare mixture models having different number of components or different models at all (for example the BYM and the proprosed model) the use of the recently introduced Deviance Information Criterion (DIC, Spiegelhalter et al. (2001)) is suggested. For a generic model with possibly vector valued parameter ξ, the *Bayesian deviance* is defined by

$$D(\xi) = -2 \log p(y|\xi) + 2 \log f(y) \tag{15}$$

where the likelihood of the current model $p(y|\xi)$ is compared to the baseline term $f(y)$ depending solely on the data y, and thus not affecting posterior inferences. For distributions within the exponential family, the similarity with the classical Deviance statistics can be strengthened by assuming $f(y)$ to be the saturated likelihood

$$f(y) = \prod_i p(y_i|\mu(\xi_i) = y_i) \tag{16}$$

where $\mu(\xi_i)$ is the expected value of $p(y_i|\xi_i)$ as a function of ξ_i. Notice that ξ is usually assumed to be the canonical parameter (for example, in the BYM model $\xi \equiv \varphi$ and $\mu(\varphi_i) = E_i \exp(\varphi_i)$): this choice is not compulsory since there is usually little difference in using a non-canonical parametrization, as for example is done with the mixture likelihood (5). Simple calculations using (16) show that the (saturated) Bayesian deviance (15) turns out to be

$$D_s(\theta) = 2\sum_i \left[y_i \log\left(\frac{y_i}{\theta_i E_i}\right) - y_i + \theta_i E_i \right] \tag{17}$$

where $\theta_i = \exp(\varphi_i)$ for the BYM model and $\theta_i = \psi_{z_i}$ for the proposed model: expression (17) can be easily implemented in the Winbugs 1.3 syntax and monitored at each sweep of the MCMC algorithm. The *Deviance Information Criterion* is defined by

$$\mathrm{DIC} = \overline{D} + p_D \tag{18}$$

where $\overline{D}$ is the posterior expectation of (17) and

$$p_D = \overline{D} - D_s\left(\widetilde{\theta}\right) \tag{19}$$

$\widetilde{\theta} = (\widetilde{\theta}_1, \ldots, \widetilde{\theta}_n)$, $\widetilde{\theta}_i = \exp[\mathrm{E}(\varphi_i|y)]$ for the BYM model and $\widetilde{\theta}_i = \mathrm{E}(\psi_{z_i}|y)$ for the proposed model; p_D is the *effective number of parameters* (see Spiegelhalter et al. (2001) for details), in that this is often less than the total number of parameters, due to the interdependences among parameters at the lowest level of the hierarchy (the likelihood) introduced by random effects specified at higher levels. Spiegelhalter et al. (2001) justify the proposed criterion by noticing that $\overline{D}$ can be considered as a posterior measure of the goodness of fit, while p_D is a penalty term measuring the complexity of the current model. It is clear that smaller values of the DIC indicate a better model.

4 A Synthetic Case Study

In this section a synthetic dataset is analyzed to assess the performance of the proposed model when the marginal distribution of the underlying risks shows some discontinuities. The chosen areal units consist of the 95 Italian departments in 1994 (notice that subdivision is no longer in use since 1995); the synthetic relative risk structure is shown in the first map of Fig. 1 (NORTH-SOUTH) and it consist of two well separated levels, $\psi_1 = 0.8$

corresponding to the lightest gray and $\psi_2 = 1.6$ to the darkest one. Counts y_i, $i = 1, \ldots, 95$ were independently simulated by the model $y_i \sim \text{Poisson}(\theta_i E_i)$ taking $\theta_i = \psi_1$ or $\theta_i = \psi_2$ according to the NORTH-SOUTH map and considering as expected cases E_i those obtained in a previous study concerning the mortality for endocrine and related disorders in Italy in 1994 (Bilancia and Pollice (2000)), scaled by 10 as to obtain a moderate number of areas having few people at risk. Due to the largely varying sizes of simulated counts (see the map COUNTS in Fig. 1) and expected cases throughout the map, it is not surprising that the Standardized Mortality Ratios $\text{SMR}_i = y_i/E_i$, $i = 1, \ldots, 95$, result in a quite unstable map (SMRs in Fig. 1) where the noise indeed dominates, though some evidence for a discontinuous N-S trend is still present.

The four following model were estimated using the simulated dataset:

- BYM-A, a standard Besag, York and Molliè model (Section 1) with $[\gamma_u] \sim \text{Gamma}(a_u, b_u)$, $[\gamma_\varepsilon] \sim \text{Gamma}(a_\varepsilon, b_\varepsilon)$, $a_u = 0.5$ and $b_u = 0.0005$ (neutral specification, see Bilancia and Pollice (2000)), $a_\varepsilon = 0.01$ and $b_\varepsilon = 0.01$;
- BYM-B, same as BYM-A except that $a_u = 0.01$ and $b_u = 0.01$ (such hyperprior setting generally results to bess less neutral and imposes a deeper regularization of the map);
- MIXCOV2-A, the proposed model with $k = 2$ components, $\alpha = 2$ and α/β equal to the median of observed SMRs, $a_u = 0.5$ and $b_u = 0.0005$;
- MIXCOV2-B, same as as MIXCOV2-A except that $a_u = 0.01$ and $b_u = 0.01$.

To assess their performance with respect to simulated risks, estimated models were ranked by means of the global Mean Squared Error, i.e.

$$\text{MSE} = \left\{ n^{-1} \sum_{i=1}^{n} \left(\theta_i - \hat{\theta}_i \right)^2 \right\}^{1/2} \tag{20}$$

where $\hat{\theta}_i = \text{E}[\exp(\varphi_i)|y]$ for BYM-A and BYM-B, whereas $\hat{\theta}_i = \text{E}[\psi_{z_i}|y]$ for MIXCOV2-A and MIXCOV2-B models; from a posterior inference point of view, models were ranked on the basis of the DIC. Results are reported in Table 1.

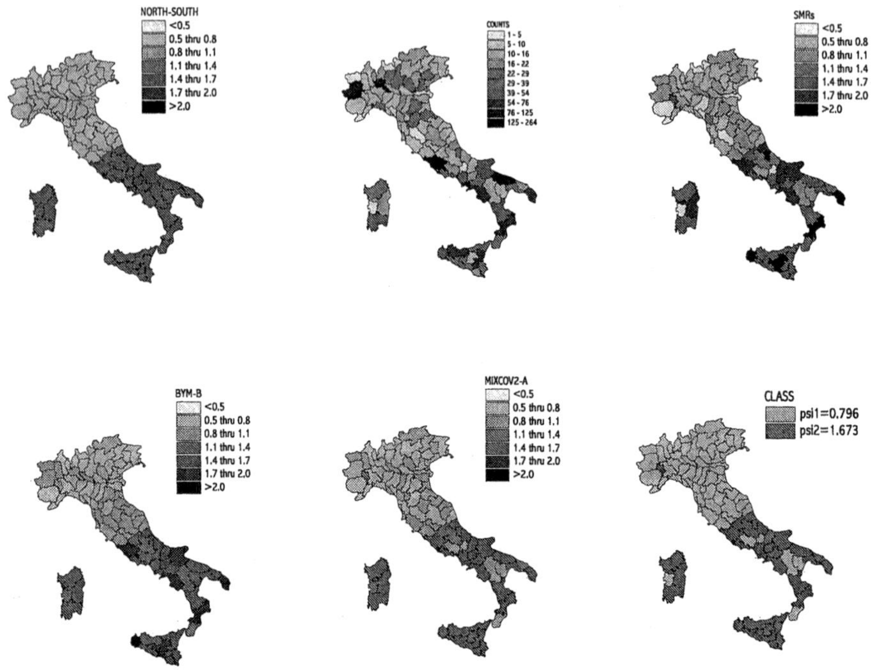

Fig. 1. NORTH-SOUTH: the synthetic risk structure with $\psi_1 = 0.8$ corresponding to the lightest gray level and $\psi_2 = 1.6$ corresponding to the darkest one. COUNTS: simulated counts, y_i, $i = 1, \ldots, 95$. SMRs: Standardized Mortality Ratios, $\text{SMR}_i = y_i/E_i$. BYM-B: posterior estimates of relative risks from the BYM-B model. MIXCOV2-A: posterior estimates of relative risks from the MIXCOV2-A model. CLASS: posterior classification of areas based on allocation probabilities (13), within the MIXCOV2-A model.

Notice that both the DIC criterion and the MSE agree in choosing the mixture model rather than the BYM; the better model according to the DIC is MIXCOV2-A, and moving to the latter from BYM-B the relative decrease of the MSE is about 18%. A careful look at maps BYM-B and MIXCOV2-A in Fig. 1 reveals that the latter produces a more reliable de-noising of the observed SMRs.

Table 1. Results of the model choice criteria for estimated models. $\overline{D}$: posterior mean of the Bayesian deviance (17). p_D: effective number of parameters. DIC: Deviance Information Criterion (18). MSE: global Mean Squared Error (20)

	$\overline{D}$	p_D	DIC	MSE
BYM-A	93.39	62.05	155.44	0.1792
BYM-B	93.21	60.19	153.40	0.1785
MIXCOV2-A	86.46	29.06	115.52	0.1462
MIXCOV2-B	87.00	29.33	116.33	0.1456

The proposed model also drastically reduces the effective number of parameters p_D, due to the intrinsically discrete structure of the space of relative risks. The last map in Fig. 1 (CLASS) shows the classification of areas made by the MIXCOV2-A model on the basis of the posterior allocation probabilities (13) (estimated by suitable Monte-Carlo averages): the agreement with the original risk structure is remarkable.

It is anyway worth noticing that in more realistic situations (characterized by a less discontinuous risk structure) the discrepancy between the performance of the proposed and the BYM model could sensibly be reduced: further extensive simulations are needed to assess this key point.

References

BESAG, J. and KOOPERBERG, C. (1995): On Conditional and Intrinsic Autoregression, *Biometrics, 82*, 733–746.

BESAG, J. YORK, J. and MOLLIÈ A. (1991): Bayesian Image Restoration with Two Application in Spatial Statistics, *Annals of the Institute of Statistical Mathematics, 43*, 1–59.

BILANCIA. M. and POLLICE, A. (2000): A Multivariate Hierarchical Bayes GLM for Spatial Ecological Analysis. In: M. Bilancia, B. Cafarelli, F. Campobasso, A. Pollice: *Alcuni Contributi Teorici ed Applicativi in tema di Statistica Spaziale.* Quaderno n. 1/2000 del Dipartimento di Scienze Statistiche, Università degli Studi di Bari, 5–35.

BRESLOW, N.E. and CLAYTON, D. (1993): Approximate Inference in Generalised Linear Mixed Models. *Journal of the American Statistical Association, 88*, 9–25.

DELLAPORTAS, P., KARLIS, D. and XELALAKI, E. (1997): Bayesian Analysis of Finite Poisson Mixtures. *Manuscript*, Department of Statistics, Athens University of Economics and Business.

FERNANDEZ, C. and GREEN, P.J. (2000): Modelling Spatially Correlated Data Via Mixtures: a Bayesian Approach. *Manuscript*, University of Bristol.

GREEN, P.J. and RICHARDSON, S. (2000): Spatially Correlated Allocation Models for Count Data. *Manuscipt*, University of Bristol.

KNORR-HELD, L. and RASSER, G. (2000): Bayesian Detection of Cluster and Discontinuities in Disease Maps. *Biometrics*, *56*, 13–21.

POLLICE, A. and BILANCIA, M. (2000): A Hierarchical Finite Mixture Model for Bayesian Classification in the Presence of Auxiliary Information. *Metron*, *58/3-4*, 109–131.

SPIEGELHALTER, D.J., THOMAS, A., BEST, N. (2000): *Winbugs Version 1.3 User manual*, MRC Biostatistics Unit, University of Cambridge.

SPIEGELHALTER, D.J., BEST, N.G, CARLIN, B.P. and VAN DER LINDE, A. (2001): Bayesian Measures of Model Complexity and Fit. *Technical Report 01/13*, Division of Biostatistics, School of Public Health, University of Minnesota.

VIALLEFONT, V., RICHARDSON, S. and GREEN, P.J. (2000): Bayesian Analysis of Poisson Mixtures. *Manuscript*, Department of Statistics, University of Bristol.

Second-order Interaction in a Trivariate Generalized Gamma Distribution *

Salvatore Bologna and Gianfranco Lovison

Dipartimento di Scienze Statistiche e Matematiche "S.Vianelli", Universitá degli Studi di Palermo, Viale delle Scienze, I-90128 Palermo, Italy (bologna@unipa.it, lovison@unipa.it)

Abstract. The concept of second- (and higher-) order interaction is widely used in categorical data analysis, where it proves useful for explaining the interdependence among three (or more) variables. Its use seems to be less common for continuous multivariate distributions, most likely owing to the predominant role of the Multivariate Normal distribution, for which any interaction involving more than two variables is necessarily zero. In this paper we explore the usefulness of a second-order interaction measure for studying the interdependence among three continuous random variables, by applying it to a trivariate Generalized Gamma distribution proposed by Bologna(2000).

1 Introduction

The concept of *interaction* is ubiquitous and somewhat ambiguous in statistics. In general terms, it is meant to denote a joint feature of a set of variables, which in some sense perturbs, or modifies, the behaviour of those variables taken separately. This is a vague definition, which needs appropriate qualifications to be formally handled. Such qualifications can be specified according to different points of view:

- **direction:** an interaction assumes a different interpretation when the relationships we are examining are directed (*asymmetric*) or undirected (*symmetric*). From this point of view, we can distinguish between
  - an *asymmetric specification*: an interaction is a *joint effect* of a set of explanatory variables on a set of response variables, which modifies the main effects the variables would

* Work financially supported by P.R.I.N. cofin. MIUR 2000 grant MM13208412_001 "Statistica nella valutazione del rischio ambientale"

have if they acted separately; this is the meaning of the term interaction in regression-type models (e.g. ANOVA, logistic regression, and the like);
  - a *symmetric specification*: an interaction is a *feature of the joint distribution* of a set of variables, which causes this distribution to depart from what it would be were the variables independent (in broad sense); this is the meaning of the term interaction in correlation-type models (e.g. log-linear models for contingency tables).
- **level:** an interaction can be measured at different levels of characterization of the distribution under examination. From this point of view, we can distinguish between
  - a *distributional interaction*, when the interaction (asymmetric or symmetric) involves the whole distribution of interest;
  - a *moment interaction*, when the interaction (asymmetric o symmetric) involves only one or a few moments of the distribution of interest
- **mathematical form:** even when the specification of what is meant by interaction is agreed upon, there is still room for arbitrariness in its mathematical definition, and therefore on how to measure it. As a consequence, in the literature additive, log-additive, multiplicative, etc. interactions have been proposed.

Probably the most complete and insightful work on the problems and subtleties of the concept of interaction is due to Darroch (see, for example Darroch, 1983). One interesting point that arises here concerns the parallelism between continuous and categorical distributions. Unlike what happens for other ideas and methods in multivariate analysis, the concept of interaction, and particularly of second-, third- or higher-order interaction, is better established and more largely used in the analysis of multivariate categorical distributions (e.g. through the use of log-linear models) than in the analysis of continuous multivariate distributions. In the sequel, we shall argue that this is due to the predominant role among the latter of the Multivariate Normal distribution, for which all interactions of order higher than one are absent.

In this paper, we define measures of distributional second-order interaction, both symmetric and asymmetric, which can be used for both continuous and categorical variables, but focus on their use for the former. In order to appreciate the potentials of analyzing interactions of order higher than one in the continuous case, it is necessary to turn to non-Normal multivariate distributions: as an example, we explore the usefulness of the proposed measures of second-order interaction applying it to a trivariate Generalized Gamma distribution proposed by Bologna(2000).

2 Building Interaction Measures for Multivariate Continuous Distributions

According to Bologna, Lovison (2001), there are basically two approaches in the literature for building up (higher order) interaction measures for multivariate continuous distributions:

1. choose a sensible interaction measure for categorical variables, and try to extend it to the multivariate continuous case (this is essentially the approach followed by Holland e Wang, 1987, in introducing their *local dependence function*);
2. define from scratch an interaction as a "departure from a reference model" and build up a measure with the property of being zero when the model holds (this is essentially the approach followed by Whittaker, 1990, in introducing his *partial derivative measures of interaction*).

2.1 Building a Second Order Symmetric Interaction (SOSI) Measure

- **The Holland and Wang Approach**

Holland e Wang(1987) proposed a measure of symmetric interaction for *bivariate* distributions extending the Cross Product Ratio used for categorical variables to the continuous case. Along this line of reasoning we choose the log Ratio of Cross Product Ratios (RCPR) as an optimal measure of second order interaction among three categorical variables. Let X, Y, Z

be three binary r.v.'s, each taking the two values {0,1}, and let $f_{X,Y,Z}(x,y,z) = \Pr\{X = x, Y = y, Z = z\}$ be their joint p.d.f.. The log Ratio of Cross Product Ratios is:

$$\log RCPR = \log \left[\frac{f_{X,Y,Z}(1,1,1) f_{X,Y,Z}(1,0,0) f_{X,Y,Z}(0,1,1) f_{X,Y,Z}(0,0,0)}{f_{X,Y,Z}(1,0,1) f_{X,Y,Z}(1,1,0) f_{X,Y,Z}(0,0,1) f_{X,Y,Z}(0,1,0)} \right]$$

The extension to the continuous case proceeds from defining an infinitesimal parallelepiped with coordinates (x, y, z) at the bottom left corner and $(x+dx, y+dy, z+dz)$ at the top right corner, calculating the continuous analog of $\log RCPR$ and finally finding the limit (for typographical reasons, only in first row of the following formula we used $f(\cdot)$ instead of $f_{X,Y,Z}(\cdot)$):

$$\lim_{\substack{dx \to 0 \\ dy \to 0 \\ dz \to 0}} \frac{\log \left[\frac{f(x+dx,y+dy,z+dz) f(x+dx,y,z) f(x,y+dy,z+dz) f(x,y,z)}{f(x+dx,y,z+dz) f(x+dx,y+dy,z) f(x,y,z+dz) f(x,y+dy,z)} \right]}{dx dy dz}$$

$$= \frac{\partial^3 \log f_{X,Y,Z}(x,y,z)}{\partial x \partial y \partial z}$$

• The Whittaker Approach

Since in this section we are focussing on the study of symmetric relationships, the distribution of interest is the joint p.d.f. $f_{X,Y,Z}(x,y,z)$, whose natural logarithm, by the factorization theorem, can be decomposed as:

$$\begin{aligned}\log f_{X,Y,Z}(x,y,z) = \theta + \log h_X(x) + \log h_Y(y) + \log h_Z(z) + \\ \log h_{X,Y}(x,y) + \log h_{X,Z}(x,z) + \\ \log h_{Y,Z}(y,z) + \log h_{X,Y,Z}(x,y,z)\end{aligned}$$

In order to build up a measure of symmetric second-order interaction by the partial derivative approach, we choose as reference model:

$$\begin{aligned}\log f_{X,Y,Z}(x,y,z) = \theta + \log h_X(x) + \log h_Y(y) + \log h_Z(z) + \\ \log h_{X,Y}(x,y) + \log h_{X,Z}(x,z) + \log h_{Y,Z}(y,z)\end{aligned}$$

which can be termed *no second-order interaction model*, since it involves functions of the pairs of variables (X, Y), (X, Z), (Y, Z), but not of the triplet (X, Y, Z). Following Whittaker(1990):

$$\frac{\partial^3 \log f_{X,Y,Z}(x,y,z)}{\partial x \partial y \partial z} = \frac{\partial^3 \log h_{X,Y,Z}(x,y,z)}{\partial x \partial y \partial z}$$

is a measure of second order symmetric interaction, since it is zero if the reference model holds and otherwise it quantifies the departure from it.

- **A Unified Measure of Second Order Symmetric Interaction**

Since the measures arrived at through the two approaches come up to be the same (a result curiously neglected in the literature), we can denote the measure by the common acronym SOSI (Second Order Symmetric Interaction):

$$\mathrm{SOSI} = \frac{\partial^3 \log f_{X,Y,Z}(x,y,z)}{\partial x \partial y \partial z} \tag{1}$$

2.2 Building a Second Order Asymmetric Interaction (SOAI) Measure

- **The Holland and Wang Approach**

The asymmetric case does not seem to have been considered by Holland and Wang. However, according to their approach, it is natural to take as an optimal measure of second order asymmetric interaction for categorical variables the Difference of Differences of conditional logits (DDCL). Denoting by $\mathrm{logit} f_{Z|X,Y}(1 \mid x,y) = \log\left[\frac{\Pr\{Z=1|x,y\}}{\Pr\{Z=0|x,y\}}\right]$ the logit of Z conditional on the pair of explanatory binary variables (X, Y), DDCL is defined as:

$$\begin{aligned} DDCL = & \left(\mathrm{logit} f_{Z|X,Y}(1 \mid 1,1) - \mathrm{logit} f_{Z|X,Y}(1 \mid 1,0)\right) \\ & -\left(\mathrm{logit} f_{Z|X,Y}(1 \mid 0,1) - \mathrm{logit} f_{Z|X,Y}(1 \mid 0,0)\right) \end{aligned}$$

The first step in the extension of DDCL to the continuous case is to define a continuous analog of the logit transformation:

$$\text{logit} f_{Z|X,Y}(z \mid x, y) = \lim_{w0} \frac{\log\left[\frac{f_{Z|X,Y}(z+dz|x,y)}{f_{Z|X,Y}(z|x,y)}\right]}{dz} =$$

$$= \frac{\partial \log f_{Z|X,Y}(z \mid x, y)}{\partial z}$$

Then, the extension of DDCL to the continuous case is easily found to be:

$$\lim_{\substack{dx \to 0 \\ dy \to 0}} \left[\left(\frac{\partial \log f_{Z|X,Y}(z \mid x+dx, y+dy)}{\partial z} - \frac{\partial \log f_{Z|X,Y}(z \mid x+dx, y)}{\partial z}\right) - \right.$$

$$\left.\left(\frac{\partial \log f_{Z|X,Y}(z \mid x, y+dy)}{\partial z} - \frac{\partial \log f_{Z|X,Y}(z \mid x, y)}{\partial z}\right)\right] / dxdy$$

$$= \frac{\partial^2}{\partial x \partial y}\left[\frac{\partial \log f_{Z|X,Y}(z \mid x, y)}{\partial z}\right]$$

• The Whittaker Approach

Since we are now focussing on the study of asymmetric relationships, the distribution of interest is the conditional p.d.f. $f_{Z|X,Y}(z \mid x, y)$, whose natural logarithm, by the factorization theorem, can be decomposed as:

$$\log f_{Z|X,Y}(z \mid x, y) = \log h_Z(z) + \log h_{X,Z}(x, z) + \log h_{Y,Z}(y, z) + \log h_{X,Y,Z}(x, y, z)$$

The reference model is here that of *absence of interactive effect of (X, Y) on Z*:

$$\log f_{Z|X,Y}(y \mid x, y) = \log h_Z(z) + \log h_{X,Z}(x, z) + \log h_{Y,Z}(y, z)$$

which postulates separate effects of X and Y on Z, but no joint effects of the pair (X, Y) on Z. Then:

$$\frac{\partial^2}{\partial x \partial y}\left[\frac{\partial \log f_{Z|X,Y}(z \mid x, y)}{\partial z}\right]$$

is a measure of second order asymmetric interaction, since it is zero if the reference model holds and otherwise it quantifies the departure from it, i.e. the intensity of the interactive effect of (X, Y) on Z.

• A Unified Measure of Second Order Asymmetric Interaction

Again, the same measure is obtained along either approach, and we can denote it by the common acronym SOAI (Second Order Asymmetric Interaction):

$$\mathrm{SOAI} = \frac{\partial^2}{\partial x \partial y}\left[\frac{\partial \log f_{Z|X,Y}(z \mid x, y)}{\partial z}\right] \tag{2}$$

Finally, notice that, although from a purely mathematical point of view the SOSI and SOAI measures coincide, it is useful to keep them distinct to stress their different (symmetric vs. asymmetric) interpretation.

3 The Normal Case

In this section, we explore the behaviour of the SOSI and SOAI measures in the case of a trivariate centered and scaled Normal random vector:

$$[X, Y, Z]' \sim \mathcal{N}_3(\boldsymbol{\mu}, \boldsymbol{\Sigma}), \boldsymbol{\mu} = \mathbf{0}, \boldsymbol{\Sigma} = \boldsymbol{P}$$

To derive the measure of symmetric interaction, it is convenient to write the trivariate Normal log-density as:

$$\log f_{X,Y,Z}(x, y, z) = \theta + \theta^X x^2 + \theta^Y y^2 + \theta^Z z^2 + \theta^{XY} xy + \theta^{XZ} xz + \theta^{YZ} yz$$

where: $\theta = -\frac{1}{2}\log(|\, \boldsymbol{P} \,|) - \frac{3}{2}\log(2\pi)$

$$\theta^X = -\frac{1}{2}\frac{(1-\rho^2_{YZ})}{|\, \boldsymbol{P} \,|} \qquad \theta^Y = -\frac{1}{2}\frac{(1-\rho^2_{XZ})}{|\, \boldsymbol{P} \,|}$$

$$\theta^Z = -\frac{1}{2}\frac{(1-\rho^2_{XY})}{|\, \boldsymbol{P} \,|} \qquad \theta^{XY} = \frac{(\rho_{XY} - \rho_{XZ}\rho_{YZ})}{|\, \boldsymbol{P} \,|}$$

$$\theta^{XZ} = \frac{(\rho_{XZ} - \rho_{XY}\rho_{YZ})}{|\, \boldsymbol{P} \,|} \qquad \theta^{YZ} = \frac{(\rho_{YZ} - \rho_{XY}\rho_{XZ})}{|\, \boldsymbol{P} \,|}$$

Clearly:

$$\text{SOSI} = \frac{\partial^3 \log f_{X,Y,Z}(x,y,z)}{\partial x \partial y \partial z} = 0$$

In order to study the asymmetric interaction, suppose (X, Y) are explanatory variables, and Z is the response variable. Then, the logit of the distribution of Z conditional on the pair (X, Y) is:

$$\text{logit} f_{Z|X,Y}(z \mid x, y) = 2\theta^Z z + \theta^{XZ} x + \theta^{YZ} y$$

where: $\theta^Z = -\frac{1}{2}\frac{(1-\rho^2_{XY})}{|\, \boldsymbol{P} \,|}$

$$\theta^{XZ} = \frac{(\rho_{XZ} - \rho_{XY}\rho_{YZ})}{|\, \boldsymbol{P} \,|} \qquad \theta^{YZ} = \frac{(\rho_{YZ} - \rho_{XY}\rho_{XZ})}{|\, \boldsymbol{P} \,|}$$

whence, obviously, we obtain:

$$\text{SOAI} = \frac{\partial^2}{\partial x \partial y}\left[\frac{\partial \log f_{Z|X,Y}(z \mid x, y)}{\partial z}\right] = 0$$

So, whether we look at it from the symmetric or the asymmetric point of view, the trivariate Normal distribution is characterized by the absence of second order interactions; moreover, it is easy to show that this conclusion extends to the multivariate Normal distribution of any dimension, which has all interactions of order higher than one equal to zero. This is hardly a surprise, since it confirms from a different perspective the well known result that the multivariate Normal distribution is completely characterized by it first two (multivariate) moments, but it is useful to recall it here, because, keeping in mind the predominant role played historically by the Normal among the continuous multivariate distributions, it may explain the scarce interest in the study of second-, and higher order, interactions in the continuous case.

4 A Non-normal Case: a Trivariate Generalized Gamma Distribution

Bologna(2000) proposed a trivariate Generalized Gamma distribution, generalizing his bivariate Gamma distribution (Bologna, 1987), by deriving the distribution of the random vector obtained by the transformation:

$$X = (T^2)^{\frac{1}{r}}, \quad Y = (U^2)^{\frac{1}{r}}, \quad Z = (V^2)^{\frac{1}{r}}; \qquad r > 0$$

when $[T, U, V]' \sim \mathcal{N}_3(\boldsymbol{\mu}, \boldsymbol{\Sigma})$.

For the sake of simplicity we consider in this paper only the special case in which:

$$\boldsymbol{\mu} = \mathbf{0}, \quad \sigma_{jj} = 1 \ \forall j\,, \quad \boldsymbol{\Sigma} = \boldsymbol{P} = (1-\rho)\boldsymbol{I} + \rho\boldsymbol{J}, \text{with } \boldsymbol{J} = \mathbf{1}\mathbf{1}'$$

i.e. the case of three centered, scaled and equicorrelated Normal r.v.'s. In this simplified case, it can be shown that:

$$\begin{aligned} f_{X,Y,Z}(x,y,z) = {} & \frac{1}{8\sqrt{2}\pi^{3/2}\sqrt{|\,\boldsymbol{P}\,|}} r^3 x^{(\frac{r}{2}-1)} y^{(\frac{r}{2}-1)} z^{(\frac{r}{2}-1)} \\ & e^{\frac{1}{2}[-\alpha(x^r+y^r+z^r)-2\tau(x^{r/2}y^{r/2}+x^{r/2}z^{r/2}+y^{r/2}z^{r/2})]} \\ & \left(1 + e^{2x^{r/2}\tau(y^{r/2}+z^{r/2})} + e^{2y^{r/2}\tau(x^{r/2}+z^{r/2})} + e^{2z^{r/2}\tau(x^{r/2}+y^{r/2})}\right) \end{aligned}$$

where: $|\boldsymbol{P}| = (1 - 3\rho^2 + 2\rho^3)$, $\alpha = \frac{1-\rho^2}{|\boldsymbol{P}|}$, $\tau = \frac{(\rho^2-\rho)}{|\boldsymbol{P}|}$.

Suppose we are interested in the study of symmetric interaction. Then, it is useful to write the log-density as:

$$\begin{aligned} \log f_{X,Y,Z}(x,y,z) = {} & \theta + \log h_X(x) + \log h_Y(y) + \log h_Z(z) + \\ & \log h_{X,Y}(x,y) + \log h_{X,Z}(x,z) + \\ & \log h_{Y,Z}(y,z) + \log h_{X,Y,Z}(x,y,z) \end{aligned}$$

where: $\theta = \log\left[\dfrac{r^3}{8\sqrt{2}\pi^{3/2}\sqrt{|\boldsymbol{P}|}}\right]$

$$\log h_X(x) = (\frac{r}{2}-1)\log x - \frac{\alpha x^r}{2} \quad \log h_Y(y) = (\frac{r}{2}-1)\log y - \frac{\alpha y^r}{2}$$
$$\log h_Z(z) = (\frac{r}{2}-1)\log z - \frac{\alpha z^r}{2} \quad \log h_{X,Y}(x,y) = -\tau x^{r/2} y^{r/2}$$
$$\log h_{X,Z}(x,z) = -\tau x^{r/2} z^{r/2} \quad \log h_{Y,Z}(y,z) = -\tau y^{r/2} z^{r/2}$$

$$\log h_{X,Y,Z}(x,y,z) =$$
$$\log\left[(1+e^{2x^{r/2}\tau(y^{r/2}+z^{r/2})} + e^{2y^{r/2}\tau(x^{r/2}+z^{r/2})} + e^{2z^{r/2}\tau(x^{r/2}+y^{r/2})}\right]$$

We get: $\text{SOSI} = \frac{\partial^3 \log h_{X,Y,Z}(x,y,z)}{\partial x \partial y \partial z} = \psi(x,y,z,r,\tau)$

where $\psi(x,y,z,r,\tau)$ is a quite complicated function of its arguments. The precise form of such a function can be found in Bologna, Lovison (2001); what is crucial here is that, unlike in the Normal case, the SOSI measure depends on r, on all of the three variables x, y, z (i.e. it is a *local* measure of interaction), and on ρ only through the reparameterization $\tau = (\rho^2 - \rho)/ \mid \boldsymbol{P} \mid$.

Figure 1 shows SOSI as a function of (x, y), for $\rho = 0.8$ (which corresponds to $\tau = -1.53846$) and nine combinations of the following values of r and z: $r = 2, 4, 6$, $z = 0.3, 1, 3$. Inspection of Figure 1 suggests some interesting conclusions:

- the three variables interact, most of the time positively (for smaller values of X, Y, Z) and less often negatively (for intermediate values of X, Y, Z);
- as X, Y, Z get large, SOSI approaches zero, i.e. they cease to interact;
- all previous patterns are accentuated as r increases.

Some insight into the previous result can be gained by considering SOSI as a measure of the variation in the first-order interaction between X and Y (conditional on Z) induced by variations in Z: where SOSI is positive (negative), increments in Z cause an increase (decrease) in the interaction between X and Y; where

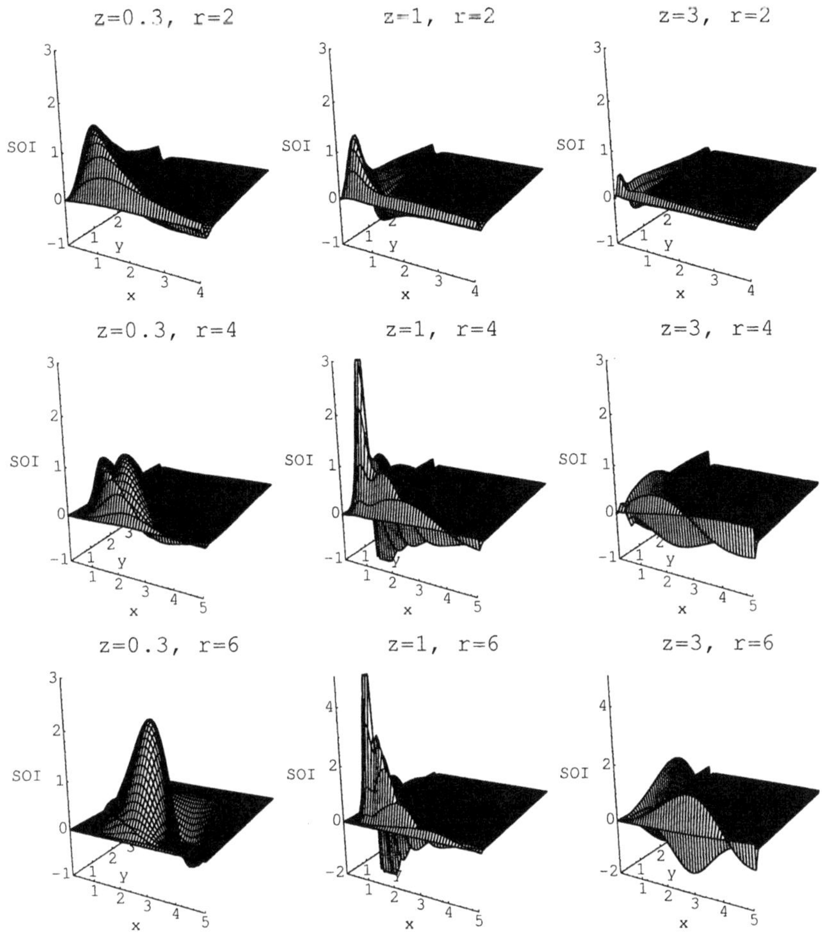

Fig. 1. The SOSI measure for various values of r and z, with $\rho = 0.8$.

SOSI is equal to, or approaches, zero, the interaction between X and Y is, or tends to be, constant with respect to Z.

This interpretation may be further helped by a visual examination of the behaviour of the conditional bivariate distributions of the pair (X, Y) given Z, with respect to variations in Z. To support such examination, Figure 2 visualizes the contour levels of the conditional distributions of the pair (X, Y) given four specified values of $Z(z = 0.3, 1, 1.3, 2)$, with $\rho = 0.8$ and $r = 4$.

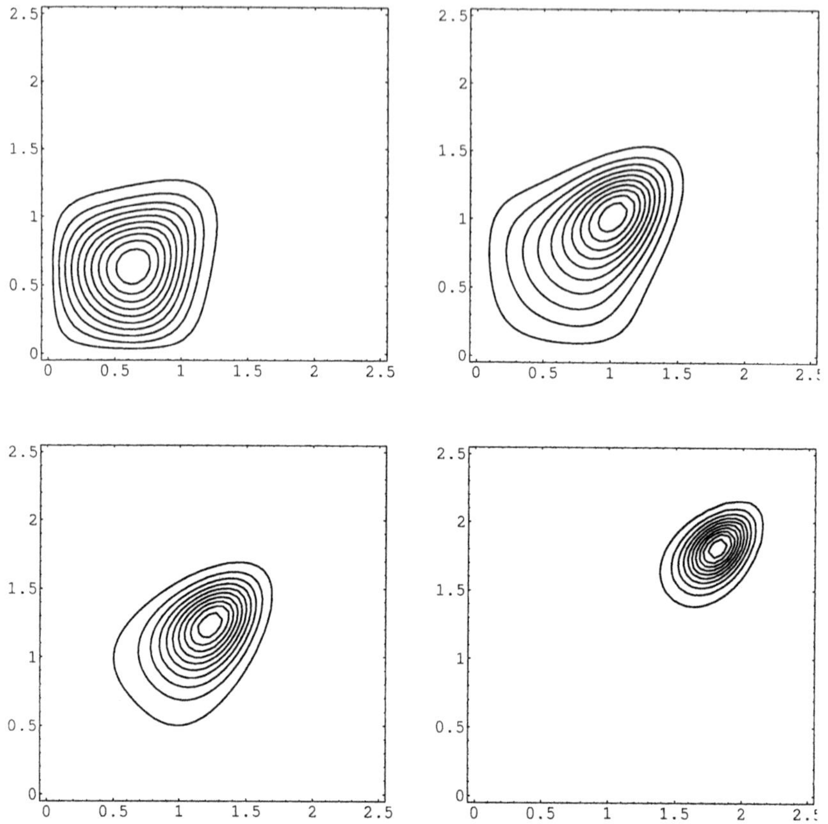

Fig. 2. Contour levels of the conditional distributions of $(X, Y)|Z$, with $\rho = 0.8$ and $r = 4$, corresponding to four specified values of Z (from top left to bottom right: $z = 0.3, 1, 1.3, 2$)

Inspection of Figure 2 provides a new interpretation of the pattern of signs in Figure 1. In fact, notice how the shape, and not only the location, of these four bivariate distributions changes as a response to changes in the values of Z, moving from a situation of positive bivariate asymmetry (for small values of Z, i.e. much less than 1), through a situation of negative bivariate asymmetry (for intermediate values of Z, i.e. values around 1), finally to a situation of approximate bivariate symmetry (for large values of Z, i.e. for values much greater than 1), which is characterized by almost elliptical contour levels.

References

BOLOGNA S. (1987): Su una distribuzione gamma generalizzata doppia. *Statistica, XLVII, 4*, 543–548.

BOLOGNA, S. (2000): Una distribuzione Gamma generalizzata multivariata. *Annali della Facoltà di Economia - Università di Palermo (to appear).*

BOLOGNA, S. and LOVISON, G. (2001): Higher-order interactions in multivariate continuous distributions. *Working paper available from the authors.*

DARROCH, J.N. (1983): Additive and multiplicative models and interactions. *Annals of Statistics, 11(3)*, 724–738.

HOLLAND, P.W. and WANG, Y.J. (1987) Dependence function for continuous bivariate densities. *Commun.Statist.-Theory Meth. 16(3)*, 863–876.

WHITTAKER, J. (1990) *Graphical Models in Applied Multivariate Statistics.* Wiley, New York.

Mortality and Air Pollution in Philadelphia: A Dynamic Generalized Linear Modelling Approach

Monica Chiogna and Carlo Gaetan

Dipartimento di Scienze Statistiche, Università di Padova, Via Battisti 241, I-35121 Padova, Italy

Abstract. In this paper, we tackle the study of the relationship between daily non accidental deaths and air pollution in the city of Philadelphia in the years 1974 – 1988. For modelling the data, we propose to make use of dynamic generalized linear models. These models allow to deal with the serial dependence and time-varying effects of the covariates. Inference is performed by using extended Kalman filter and smoother.

1 Introduction and Motivation

Various applied fields, like environmental statistics or environmental epidemiology, deal with time series data in the form of discrete or non-normal outcomes. In environmental epidemiology, a key problem is the role of serial correlation in the modelling framework. Serial correlation on the response is due to different sources of causes. First of all, outcomes depend on serially correlated explanatory variables, so that the time series structure of the covariates imparts a highly structured pattern of interdependence on the response. Then, the effect of the explanatory variables on the outcomes usually lasts some time; for example, the effect of a high pollution event on population health spreads over some days, although the effect's mechanisms is unfortunately unknown. This, again, depends partly on the serial correlation of the pollutants, partly on the natural response mechanism of the human body to exposure to toxic agents.

A natural way to deal with such issues would be to develop association models in which the dependence structure within the explanatory variables and between covariates and response is correctly accounted for. However, in most cases neither the dependence mechanism on the explanatory variables nor an adequate knowledge of the physical association between response and covariates is known, so that an explicit probabilistic model of association is rarely available.

In the environmental epidemiology literature, the usual modelling strategy (see Brumback *et al.* (2000), for up to date references) is based on estimating proper Generalized Linear Model (GLM) or Generalized Additive Models (GAM) with the assumption of independent outcomes. The temporal behavior is controlled by mean of a cyclic function inserted in the regressor term. In the diagnostic phase, checks for the presence of serial correlation in the residuals are performed. In general, residuals' serial correlation happens when the model for the temporal component is inadequate to pick up all the fluctuations of the underlying outcome behavior. In these cases, a more accurate modeling of the temporal component can solve the problem. If this is not the case, the modeling strategy can be extended to account for serially correlated terms.

In the literature, two main approaches can be followed to add autocorrelation to the standard GLM or GAM setting: either a latent autocorrelated time series error is assumed for the model (generalized linear/additive model with time series error), which means that correlation between two subsequent outcomes is a known function of the marginal means of the outcomes and perhaps of some additional parameters, or correlation is inserted into the model by making the current outcome explicitly depend on past outcomes (transitional generalized linear/additive model).

Various difficulties are related to these extensions, like computational difficulties, model checking, etc. In this paper we propose to employ a different modelling strategy based on Dynamic Generalized Linear Models (DGLMs) (Fahrmeir and Tutz (1994)), which extends the first above mentioned approach. We apply this modelling framework to the analysis of daily death counts in Philadelphia (Kim *et al.* (1999)). The counts are modelled by a Poisson distribution having mean driven by a latent Markov process. In this setting, serial dependence is added to the model structure by making use of random coefficients supplemented by prior distributions adequate to take autocorrelation into account, such as for example random walks. This implies that time-varying covariates are allowed to enter the model not only via the conditional mean but also linked to the latent process. In our view, this framework allows to more neatly explore the dependence structure of the data. Estimation is performed by extended Kalman filter and smoother.

The outline of this paper is as follows. In Section 2 we introduce some features of our modelling approach, sketching also the inference procedure. Section 3 briefly describes the Philadelphia data set. Finally, Section 4 contains the results and some concluding remarks.

2 Dynamic Generalized Linear Models

Consider a series of counts $\{Y_t\}$ recorded at equally spaced times $t = 1, \ldots, T$ along with a vector of covariates $\boldsymbol{x}_t \in R^p$. Assume that it is reasonable to divide the vector of covariates into two components $\boldsymbol{x}_t = (\boldsymbol{x}'_{1t}, \boldsymbol{x}'_{2t})'$, where the first component, $\boldsymbol{x}_{1t}$, includes covariates which we expect to contribute to the 'nucleus' of the underlying mean tendency of the counts and the second component, $\boldsymbol{x}_{2t}$, includes perturbing factors, whose influence can be thought of as being constant over time. If the central tendency of the counts may be thought of as resulting from the influence of both types of covariates, then it is reasonable to model the effect of $\boldsymbol{x}_{1t}$ by a univariate latent process $\phi_t = \phi_t(\boldsymbol{x}_{1t})$ and to fix the effects of $\boldsymbol{x}_{2t}$. Therefore, we assume that the conditional distribution of Y_t given ϕ_t follows a Poisson distribution

$$Y_t|\phi_t \sim \mathcal{P}(\exp\{\phi_t + \boldsymbol{x}'_{2t}\boldsymbol{\gamma}\})$$

with $\boldsymbol{\gamma}$ representing the fixed regression coefficients for $\boldsymbol{x}_{2t}$. For the latent process ϕ_t we consider the following specification:

$$\begin{aligned}
\phi_t &= (\omega_t + \boldsymbol{x}'_{1t}\boldsymbol{\beta}_t)\\
\omega_t &= 2\omega_{t-1} - \omega_{t-2} + \delta_t \quad \text{with} \quad \delta_t \sim N(\mu_\delta, \sigma^2_\delta).\\
\boldsymbol{\beta}_t &= \boldsymbol{\beta}_{t-1} + \boldsymbol{\xi}_t \quad \text{with} \quad \boldsymbol{\xi}_t \sim N(\boldsymbol{\mu}_\xi, \Sigma_\xi).
\end{aligned}$$

In this formulation ω_t is a discrete-time analog of a continuous-time cubic spline so that the model appears as a semi-parametric model. Also the coefficients of the long-term explanatory variables move dynamically in time. This is a crucial aspect of the model. Firstly, it is possible to grasp long term changes in the effects of the covariates. Secondly, the dynamics on the coefficients allows to capture the "carry-over" effect of the covariates which usually causes the effect at time t to be influenced by covariates at previous times. The previous setting falls into the general framework of Dynamic Generalized Linear Model (DGLM) (Fahrmeir and Tutz (1994)). The latent process ϕ_t and the parameters $\boldsymbol{\gamma}$ can be cast in the transition model

$$\boldsymbol{\alpha}_t = F\boldsymbol{\alpha}_{t-1} + \boldsymbol{\varepsilon}_t, \tag{1}$$

with $\boldsymbol{\varepsilon}_t \sim \mathcal{N}(\mathbf{0}, Q)$, $\boldsymbol{\alpha}_0 \sim \mathcal{N}(\boldsymbol{a}_0, P_0)$ and $\boldsymbol{a}_0$, P_0, Q, F hyperparameters. The observation model is given by

$$Y_t|\phi_t \sim \mathcal{P}(\exp\{\boldsymbol{z}_t'\boldsymbol{\alpha}_t\}). \tag{2}$$

To exemplify, assume for convenience that $\boldsymbol{x}_t = (x_{1t}, x_{2t})'$. Then,

$$\boldsymbol{\alpha}_t = \begin{pmatrix} \omega_t \\ \omega_{t-1} \\ \beta_t \\ \gamma \end{pmatrix} = \begin{pmatrix} 2 & -1 & 0 & 0 \\ 1 & 0 & 0 & 0 \\ 0 & 0 & 1 & 0 \\ 0 & 0 & 0 & 1 \end{pmatrix} \begin{pmatrix} \omega_{t-1} \\ \omega_{t-2} \\ \beta_{t-1} \\ \gamma \end{pmatrix} + \begin{pmatrix} \delta_t \\ 0 \\ \xi_t \\ 0 \end{pmatrix} = F\boldsymbol{\alpha}_{t-1} + \boldsymbol{\varepsilon}_t,$$

where $\boldsymbol{\varepsilon}_t \sim N(\mathbf{0}, \mathrm{diag}(\sigma_\delta^2, 0, \sigma_\xi^2, 0))$ and $\boldsymbol{z}_t = (1, 0, x_{1t}, x_{2t})'$.

2.1 Moment Structure

We will explore the nature of the serial correlation implied on the counts by the dynamical structure. Consider first the mean, $\boldsymbol{a}_t$, the variance, P_t, and the autocovariance function, $P_{t,t+h}$, of $\boldsymbol{\alpha}_t$. By recursion, it is possible to show that:

$$\boldsymbol{a}_t = F^t \boldsymbol{a}_0,$$

with $F^t = F \cdots F$, t times and $F^0 = I$. Analogously:

$$P_t = \sum_{k=0}^{t-1} F^k Q F'^k + F^k P_0 F'^k$$

and

$$Cov(\boldsymbol{\alpha}_{t+h}, \boldsymbol{\alpha}_t) = F^h P_t.$$

Therefore, $\boldsymbol{\alpha}_t \sim \mathcal{N}(\boldsymbol{a}_t, P_t)$, which yields for the latent process:

$$\boldsymbol{z}_t' \boldsymbol{\alpha}_t \sim \mathcal{N}(\boldsymbol{z}_t' \boldsymbol{a}_t, \boldsymbol{z}_t' P_t \boldsymbol{z}_t)$$

and

$$Cov(\boldsymbol{z}_{t+h}' \boldsymbol{\alpha}_{t+h}, \boldsymbol{z}_t' \boldsymbol{\alpha}_t) = \boldsymbol{z}_{t+h}' F^h P_t \boldsymbol{z}_t.$$

We now turn to the moment structure of the observed counts, for which we have:

$$\mu_t = E(E(Y_t|\boldsymbol{\alpha}_t)) = E(\exp(\boldsymbol{z}_t' \boldsymbol{\alpha}_t)),$$

$$\sigma_t^2 = E(Var(Y_t|\boldsymbol{\alpha}_t)) + Var(E(Y_t|\boldsymbol{\alpha}_t)) =$$

$$= E(\exp(\boldsymbol{z}_t' \boldsymbol{\alpha}_t)) + Var(\exp(\boldsymbol{z}_t' \boldsymbol{\alpha}_t)).$$

This yields

$$\mu_t = \exp(\boldsymbol{z}_t' a_t + \boldsymbol{z}_t' P_t \boldsymbol{z}_t / 2)$$

$$\sigma_t^2 = \mu_t^2 (\exp(\boldsymbol{z}_t' P_t \boldsymbol{z}_t) - 1)$$

$$Cov(Y_{t+h}, Y_t) = \mu_{t+h} \mu_t (\exp(\boldsymbol{z}_{t+h}' F^h P_t \boldsymbol{z}_t) - 1),$$

from which we derive that:

$$Corr(Y_{t+h}, Y_t) = \frac{\exp(\boldsymbol{z}_{t+h}' F^h P_t \boldsymbol{z}_t) - 1}{\sqrt{(\exp(\boldsymbol{z}_{t+h}' P_{t+h} \boldsymbol{z}_{t+h}) - 1)(\exp(\boldsymbol{z}_t' P_t \boldsymbol{z}_t) - 1)}}.$$

Previous results show that this model setting allows to incorporate non stationary second-order processes.

2.2 Inference

DGLMs have two unknown quantities: the state vector $\boldsymbol{\alpha}_t$ and the hyperparameters. We summarize the hyperparameters in the vector $\boldsymbol{\lambda}$, we assume for the moment $\boldsymbol{\lambda}$ fixed and known and we are interested in dealing first with $\boldsymbol{\alpha}_t$. Let $\boldsymbol{\alpha}_t^* = (\boldsymbol{\alpha}_0', \ldots, \boldsymbol{\alpha}_t')'$ and $Y_t^* = (Y_1, \ldots, Y_t)'$, the conditional distribution of $\boldsymbol{\alpha}_T^*$ given the observations

$$p(\boldsymbol{\alpha}_T^* | Y_T^*) \propto \prod_{t=1}^{T} p(Y_t|\boldsymbol{\alpha}_t) \prod_{t=1}^{T} p(\boldsymbol{\alpha}_t|\boldsymbol{\alpha}_{t-1}) p(\boldsymbol{\alpha}_0) \tag{3}$$

is non-normal. Note that in this case the conditional means and the conditional modes are not equivalent. Due to the complicated form of the conditional distributions involved, inference requires some approximation. Simulation based estimation, in particular MCMC methods, has been proposed for dealing with this problem (see Shephard and Pitt (1997), among others). In this paper we have chosen the approach proposed by Fahrmeir and Tutz (1994) and we consider (3) as a penalized likelihood avoiding a Bayesian interpretation. More precisely, taking the logarithm of the conditional density, $PL(\boldsymbol{\alpha}_T^*)$, this yields to

$$PL(\boldsymbol{\alpha}_T^*) = \text{const} + \sum_{t=1}^{T} l(\boldsymbol{\alpha}_t) - \frac{1}{2}(\boldsymbol{\alpha}_0 - \boldsymbol{a}_0)' P_0^{-1} (\boldsymbol{\alpha}_0 - \boldsymbol{a}_0)$$
$$- \frac{1}{2} \sum_{t=1}^{T} (\boldsymbol{\alpha}_t - F\boldsymbol{\alpha}_{t-1})' Q^{-1} (\boldsymbol{\alpha}_t - F\boldsymbol{\alpha}_{t-1}). \tag{4}$$

with $l(\boldsymbol{\alpha}_t) = \log p(Y_t|\boldsymbol{\alpha}_t)$. The function (4) is a penalized log-likelihood criterion so the conditional modes $\hat{\boldsymbol{\alpha}}_T^* = \text{argmax}_{\boldsymbol{\alpha}_T^*}$ $p(\boldsymbol{\alpha}_T^*|Y_T^*)$, are maximizers of $PL(\boldsymbol{\alpha}_T^*)$. Algorithmic solutions can be efficiently obtained by iterative Kalman filtering and smoothing (Fahrmeir and Wagenpfeil (1997)). The hyperparameters $\boldsymbol{\lambda}$ can be interpreted as smoothing parameters and these can be selected as minimizers of a generalized cross-validation criterion.

Diagnostics deserve some care: particularly Pearson-type residuals loose the usual properties. For more appropriate inference we

Table 1. Summary statistics for Philadelphia: Mortality age 65+, temperature and dewpoint (^oF), TSP (μg/m^3).

Variable	Mean	SD	25%	50%	75%
Mort	31.5	6.4	26.0	31.0	38.0
Temp	54.3	17.8	40.0	55.3	70.3
Dew	42.3	19.1	27.8	43.5	58.8
TSP	67.3	26.9	47.5	63.0	72.0

use the P-scores as suggested by Früwirth-Schnatter (1996), and we perform standard diagnostic tools for generalized linear models as well as dynamic linear models for detecting model's inadequacies, if any.

3 Data

The first substantial analysis of a U.S. epidemiological time series data set was by Schwartz and Dockery (1992), using data from Philadelphia and it was quickly followed by a number of studies of other cities. The main Philadelphia data set used by the researchers consisted of 14 years of daily deaths data (1974–1988) with associated measurements of temperature and dewpoint, i.e. the two meteorological variables which are believed to be the most important confounders, and five pollutants: total suspended particulate (TSP), sulphur dioxide (SO2), nitrogen dioxide (NO2), carbon monoxide (CO) and ozone (O3).

In the present analysis, we have used the same data set as in Kim *et al.* (1999), and we have focused on the analysis of TSP effects on deaths in the population aged 65 and over, since this is the group most at risk. Table 1 includes summary statistics for the daily air pollutant concentration, key meteorology and death counts recorded during the study period.

Figure 1, which plots weekly deaths counts along with a non parametric estimate of the temporal behaviour, shows that there is an irregular seasonal effect, which cannot be explained solely through the dependence of deaths on either meteorology or pollution and need to be properly modelled. Moreover, extensive preliminary analyses show that deaths decrease against both temperature and dewpoint until a threshold value is reached, around

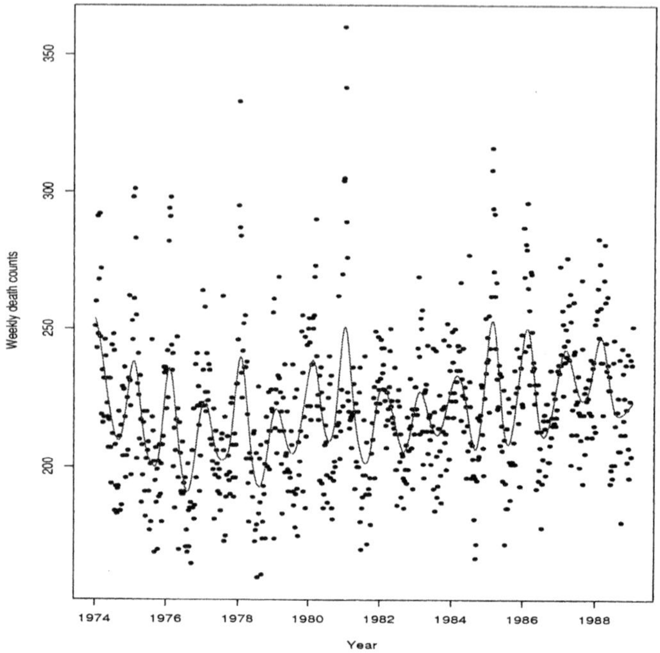

Fig. 1. Weekly deaths counts for the years of study (1974–1988) along with a nonparametric estimate of the temporal behaviour (solid line). Dotted vertical lines mark the beginning of each calendar year.

75° F for temperature and 60° F for dewpoint, after which deaths start to increase with increasing temperature or dewpoint. Similar analyses for TSP show a general increase in deaths with pollutant, although it is questionable whether there is any real effect below about 100 $\mu g/m^3$. To develop the models, we take advantage of the exploratory analysis and of previous studies (see Kim *et al.* (1999)) to construct and select sound covariates, although we do not aim at building models which are strictly comparable with the models already published.

4 Results and Conclusions

Our model building strategy started from the simplest models, i.e. models including the pollutant and the most relevant meteo-

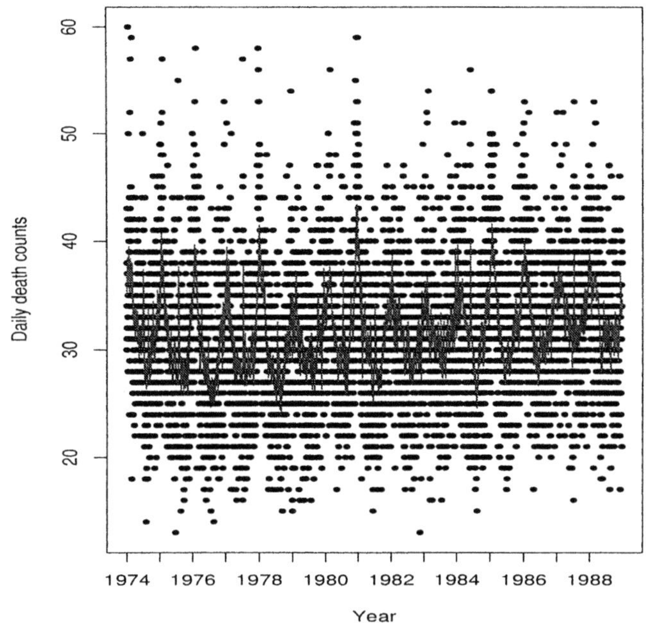

Fig. 2. Time series of the observed counts along with fitted values.

rological variables. If variables resulted to be not significant, i.e. the corresponding time-varying confidence bands always included zero, they were removed from the models.

As we assumed that the counts reflected an underlying tendency of the severe air pollution events, combined with adverse meteorological conditions, to cause non-accidental death, we included in the $\boldsymbol{x}_{1t}$ component vector those covariates which measured exposure to pollution and meteorological conditions.

Long term trends and seasonal fluctuations were controlled by making use of a discrete version of a spline component modelled by a second order random walk. Controlling for weather was achieved by including a time varying coefficient for the mean temperature of the previous 3 days less than 80^0 F and a fixed effect for temperature above the same cutoff. Moreover, a time varying effect for the mean dewpoint temperature of the previous 3 days was also introduced.

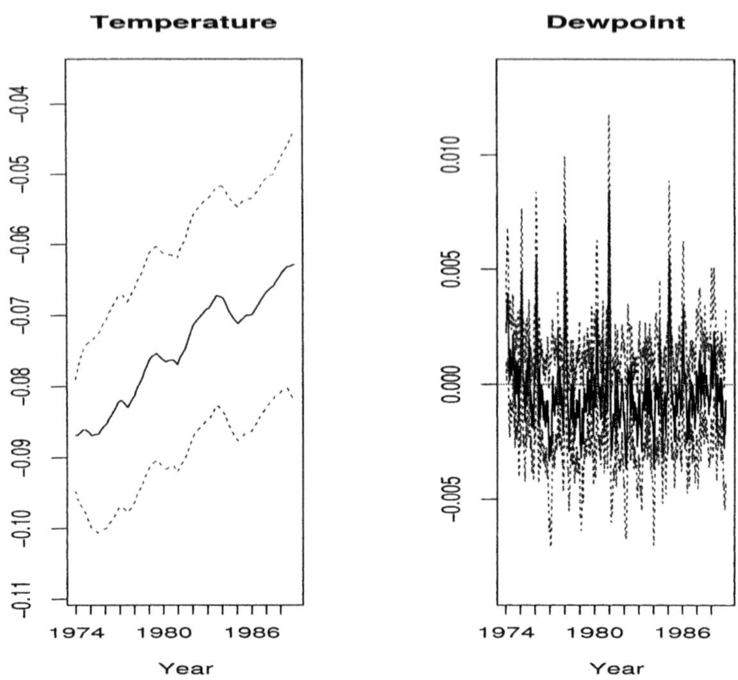

Fig. 3. Estimated trajectories for the coefficients of temperature below 80 ^oF and of the dewpoint temperature with point-wise 95% confidence intervals.

All analyses were carried out using R functions and C routines written by ourselves.

Figure 2 shows the time series of the observed counts along with the fitted values.

Results showed a significant and positive effect of TSP on health. The pollutant's effect resulted to be fixed, i.e. the estimated coefficient variance was near to zero, meaning that the three day lagged measure was sufficient to capture the carry-over effect of the pollutant.

Figure 3 shows the estimated trajectories for the coefficients of the temperature below 80 ^oF and of the dewpoint temperature and highlights the time varying effects of the variables. The temperature effect appears to be significant over the study period, whereas controversial appears the interpretation of the dewpoint temperature. It appears that most of the dynamic behaviour of the time series is captured by dewpoint temperature so that it becomes difficult to assess significance of the variable and the strength of its effect.

The aim of our work was to explore the extent to which the DGLM setting allows to improve modelling procedures for epidemiological time series studies, where a complex dependence structure among variables is observed. Results seem satisfying in terms of flexibility of the modelling approach and of parsimony, where parsimony is to be intended with respect to the number of explanatory and confounding variables that need to be included in the model and not with respect to the number of parameters to be estimated.

References

BRUMBACK, B.A., RYAN, L.M., SCHWARTZ, J.D., NEAS, L.M., STARK, P.C., BURGE, H. (2000): Transitional regression models with application to environmental time series. *Journal of the American Statistical Association, 95*, 16–27.

FAHRMEIR, L. and TUTZ, G. (1994): *Multivariate Statistical Modelling Based on Generalized Linear Models.* Springer-Verlag, New York.

FAHRMEIR, L. and WAGENPFEIL, S. (1997): Penalized likelihood estimation and iterative Kalman smoothing for non-gaussian dynamic regression models. *Computational Statistics and Data Analysis, 24*, 295-320.

FRÜWIRTH-SCHNATTER, S. (1996): Recursive residuals and model diagnostics for normal and non-normal state space models. *Environmental and Ecological Statistics, 3*, 291-309.

KIM Y., SPITZNER D., ZHANG Z., SMITH R.L. and FUENTES M. (1999): Accounting for multiple pollutants in pollution-mortality studies. In: *Proceedings of the ASA - Biometrics Section*, 1-10.

SCHWARTZ, J. and DOCKERY, D.W. (1992): Increased mortality in Philadelphia associated with daily air pollution concentrations. *A. Rev. Respr. Dis., 145*, 600-604.

SHEPAHRD, N. and PITT, M.K. (1997): Likelihood analysis of non-gaussian measurement time series. *Biometrika, 84*, 653-667.

The Multivariate Adaptive Sampling for Estimating the Diversity in Biological Populations

Stefano A. Gattone and Tonio Di Battista

Dipartimento di Metodi Quantitativi e Teoria Economica Univ. "G. D'Annunzio"
V.le Pindaro, 42, 65127 Pescara

Abstract. In environmental surveys, due to environment fluctuations, the population could be spread on a wide study area so that the population density is low but many units are concentrated inside small regions. In this case, we say that the biological community is clustered. In this setting, we propose the multivariate adaptive sampling to estimate biological diversity. In particular, in order to describe the diversity of a biological population, we consider several distribution models for the abundance vector. By means of a suitable simulation study, limits and advantages of using adaptive sampling designs are emphasized for each abundance model considered.

1 Introduction

Statistical Ecology encompasses numerous quantitative methodologies that deal with the exploration of patterns in biological communities. In some cases the aim of the study is to test hypotheses about the underlying structure of the ecological community. Species abundance patterns relate to the organization of a community, including how coexisting species utilize common resources such food or space. Ecologists have developed several hypotheses in an attempt to explain these abundance species interactions. At this purpose, the concept of diversity has been used as a method for characterizing the structure of species abundance in a community. Theoretical distributions like the geometric distribution, the brocken-stick distribution or the lognormal distribution have been found and related to important aspects of diversity such as the Dominance or the Equitability.

In many environmental and natural phenomena, the population investigated is a collection of objects within a particular study area. Recently, Barabesi and Fattorini (1998) and Di Battista

(2002) have discussed several sampling techniques to estimate the species abundance of biological populations. In this paper we propose a sampling strategy based on Multivariate Adaptive Cluster Sampling (MACS) (Thompson (1993), Seber and Thompson (1994)) in order to estimate the abundance vector when a clustered pattern is suspected in a study area. Generally, when Adaptive Cluster Sampling is adopted with rare and patchy populations, the resulting sample density estimates have a lower variance than those from Simple Random Sampling (SRS). Ecologists have shown interest in this sampling technique as many of the populations in ecology have a patchy distribution (Andrew and Mapstone (1987)) and it is very rare that an ecological population is randomly distributed. By means of the abundance vector estimation we proceed in testing hypotheses about Species Abundances and find the statistical model that best fit the data. Several simulations were performed in order to identify and evaluate the factors that influence the gain of adaptive cluster sampling in comparison to simple random sampling.
The paper is organized as follows: in Section 2 we develop a procedure to estimate the abundance vector of a biological population by using adaptive cluster sampling. In Section 3 we introduce some of the most commonly used species abundance model. In Section 4, in order to stress pro and cons of using adaptive sampling designs, a thorough simulation study is performed. Results and some final conclusion are given in Section 5.

2 Abundance Vector Estimation

The diversity of a biological population (e.g. diversity and/or dominance indexes) are expressed as function, say $g(\mathbf{T})$, of the unknown abundance vector $\mathbf{T} = (T_1, T_2, ..., T_s)$, where the j-th element of the vector $\mathbf{T}$, represents the number of units belonging to each biological species ($j = 1, 2, ..., s$). The problem lies in adopting a sample design to estimate the abundance vector $\mathbf{T}$. In the literature several sampling designs have been proposed. However, when the population of the biological species can be represented by clustered points in a wide study area, the simple random sampling estimators are characterized by a large variance. Moreover, multivariate considerations in conventional sampling designs involve question of survey design. Indeed, the conventional advice that, when each variable has a sporadic distribution concentrated in a small part of the study region, a different sampling design is needed for each variable (Cochran (1977) p. 77) may not need to be adhered when an adaptive sampling design is used. In fact, with Adaptive Cluster Sampling, the course of the survey depends on the observed values, hence it is possible to accommodate in a single survey the sporadic distribution of several variables of interest. The problem can be viewed in terms of multivariate approach, where the abundances can be considered as s different variables of interest Y_j, $(j = 1, 2, ..., s)$. In this context, for each area unit $(u_1, u_2, ..., u_N)$ labelled by (1,2,...,N), the value of the j-th variable is denoted by y_{ij}, $(i = 1, 2, ..., N)$ and the population matrix may be denoted with $\mathbf{Y}$. In the fixed population view, $\mathbf{Y}$ is considered a fixed set of unknown constants and a sample of size n, drawn from the population units, represents a sequence of n labels. In adaptive sampling, each unit is defined to have a neighborhood of other units associated with it. For example, suppose the population is composed of two subpopulation:

$$P_M = \{y : y_i \in C, i = 1, 2, ..., N\}$$

and

$$P_{N-M} = \{y : y_i \ni C, i = 1, 2, ..., N\}$$

with $C \in \mathrm{R}$ a previously fixed condition. In the multivariate case, the condition is specified by a region C in s-dimensional space. So, if a member of P_M is selected to be in the initial sample, then its neighborhood is added to the sample and if any member of the neighborhood is also in P_M, then its neighborhood is also selected. Adaptive sampling stops when no more neighbors are members of P_M and the final sample consists of the initial sample and all adaptively sampled units. At the end of the procedure for each initial unit we have a corresponding set of units of P_M which we call "*network*". Note that, by symmetry of the definition of neighborhood, if any unit in a network is selected to be in the initial sample, then every member of the network is adaptively sampled and is selected as part of the final sample. Now suppose an adaptive sampling design is performed. Let us denote by $g = (i_1, i_2, ..., i_L)$ a collection of indexes drawn from $(1, 2, ..., L)$ networks, $g \in G$, where G is the set of all possible samples of networks. Then, a design unbiased estimator of the abundance T_j is

$$\hat{T}_j = \frac{N}{n} \sum_{i \in g} w_{ij} = \frac{N}{n} \sum_{l=1}^{L} Z_l w_{lj} \tag{1}$$

where

$$w_{lj} = M_l^{-1} \sum_{r \in A_l} y_{rj}$$

is the average of the l-th network ($l = 1, 2, ..., L$) and $Z_l \approx Bin(n, M_l/N)$.
The abundance estimator $\hat{T}_j$, ($j = 1, 2, ..., s$) represents an unbiased and consistent estimator of the population counterpart (Thompson, 1996). The total estimator of the population is $\hat{T} = \mathbf{1}^t \hat{\mathbf{T}}$. In addition, a straightforward application of the Central Limit Theorem ensures that the abundance vector estimator $\hat{\mathbf{T}} = (\hat{T}_1, \hat{T}_2, ..., \hat{T}_s)^t$ is normally distributed.
We point out that the estimator (1) does not depend on the condition C as it simply determines the network structure before any sampling take place.

3 Some Species Abundance Distributions

The species diversity of a biological population has always been object of environmental studies. Detailed reviews of statistical methods to analyze this problem can be found in Patil and Taillie (1982) and Gove et al. (1994).
Given a community, an assemblage of organism living together in a particular environment, ecologists attempt to answer a the following question: how do species evolve and relate to one another? The simplest measure of diversity is to count the number of species: species richness. This approach does not take species abundance

into account. Many diversity indices have been developed that incorporate both species richness and the degree of evenness with which individuals are distributed among species. There is no index that is excellent across discriminant ability (how capable they are of detecting subtle differences between sites), sensibility to sample size and to common or rare species. Therefore, it is not possible to recommend a single index as superior to all others and choosing the appropriate index depends on what sort of question is being asked. Moreover, descriptions of whole communities by one statistic of diversity run the risk or loosing much valuable information. As a matter of fact, a more complete picture is obtained by examining the species abundance pattern amenable to a more meaningful analysis. Several models to describe the patterns of a biological community have been used (Pielou (1975)). Assuming that the total number of species is known, we consider the models commonly discussed in the ecological literature, all referring to how the total niche of a community represented by a stick should be broken into pieces. Geometric model, dominance model (May (1975)), is typically adopted when one assumes hypothesis that each species time to time consumes an amount, say k, of the resource. In this context the abundance for the j-th species is given by

$$T_j^G = \mathbf{T}k(1-k)^{j-1}. \tag{2}$$

In contrast in the Brocken-stick model, *equitability* (Mac Arthur (1957)), the species simultaneously consume an amount of the resource implying random partitioning of resources, thus

$$T_j^{BS} = \mathbf{T}\frac{1}{s}\sum_{r=1}^{j}\frac{1}{s-r+1} \tag{3}$$

Finally, the lognormal model, *diversity model* (Preston (1948)), is more common for communities rich in species. One ecological explanation for the lognormal distribution refers to the brocken stick model, but the stick is broken sequentially and non instantaneously. The model can be specified by

$$S_{(R)} = S_0 \exp(-a^2R^2) \tag{4}$$

where $a = \frac{1}{\sqrt{2\sigma}}$,$R = \log_2(I_i/I_0)$, σ is the standard error of the distribution and S_0 is the number of species in the modal interval. Following Preston's convention of expressing abundance, I_i is the species abundance in the i-th interval and I_0 is the species abundance in the modal interval. $S_{(R)}$ thus represents the theoretical number of species in the R-th interval using the log normal model. Note that abundance intervals are expressed in a logarithmic scale.

We point out that the distribution of resources is most equitable (high diversity) in the brocken stick model, less equitable (medium diversity) in the lognormal model and less equitable (low diversity) in the geometric series. These models span over the whole range of possible species-abundance patterns. However, unless there are many species in the samples, it is often very difficult to distinguish a brocken stick from a lognormal model (Wilson (1993)).

4 Method: Simulation Data and Model Fitting

Simulation experiments were performed in order to identify and evaluate factors that influence the relative gain of MACS to SRS. Monte Carlo simulation techniques are used to generate 36 populations by the "modified Thomas process" of Diggle et al. (1976). The study area is represented by a study site (70x70 unit) in which parents are randomly distributed, each parent then being surrounded by a clump of offsprings. A parent coordinate pair (a, b), $0 \leq a, b \leq 1$, is selected by choosing two random values from the uniform distribution. For the j-th species, the number of offsprings for a parent is generated as a Poisson variate with mean λ_j. The average clump size for the j-th species is $\lambda_j + 1$, i.e. the number of offsprings plus the parent. Finally, the number of parents for the j-th species is calculated as $p_j = \frac{T_j}{(\lambda_j+1)}$.

The offsprings are spatially distributed about their parents at a radial distance selected from an Exponential distribution with mean r_j and at a random angle uniformly distributed between 0 and 360. The sampling area was defined as the central area of 50x50 units within the site, to allow for edge effects from the modified Thomas process. The model was used to generate 36 populations models using a $3 \times 2 \times 3 \times 2$ factorial design with the factor levels being:

- the total of the population, $T = \sum_{j=1}^{s} T_j$, was 100, 500 and 2000.
- the number of parents, p_j, was 5 and 15.
- the mean distance of offsprings from each parent, r_j , was 1, 2 and 4.
- the species abundance distribution were the Geometric model with $k = 0.6$ and the Brocken stick model both with s, the number of species equal to 7.

The first three factors cover the range of natural spatial patterns that may be encountered in most ecological communities, from rare and highly clustered population to common and randomly distributed population. The populations were each sampled 2000 times using MACS with initial sample $n = 15$ and estimates $\hat{\mathbf{T}}_{macs}$ of the abundance vector $\mathbf{T}$ were obtained using the estimator (1). The aggregative condition was defined by the region

$$C = \{y_{ij} : y_{ij} > 0, \forall i, j\}.$$

In order to evaluate the gain obtained using the adaptive strategy respect to the simple random sampling, conventional estimates $\hat{\mathbf{T}}_{srs}$ of $\mathbf{T}$ were evaluated using SRS with sample size equal to the expected sample size $E(v)$ under the adaptive design (Thompson (1996) p. 111). Finally, we proceeded in fitting the estimated abundances vector with the theoretical abundances distributions. In particular, it is worthwhile to note that ecologists, in practice, do not judge the appropriateness of models by absolute fit, but rather by whether they give a better fit than other models. Therefore, once we have the estimates $\hat{\mathbf{T}}_{macs}$ and $\hat{\mathbf{T}}_{srs}$, Dominance and Equitability models were then fitted to these abundance using the method of Wilson (1991), i.e. the sum of squared Euclidean distance between model and observation on a logarithmic scale:

$$D_{mod} = \sum_{j=1}^{s} [\log T_{obs(j)} - \log T_{mod(j)}]^2 \tag{5}$$

where $T_{obs(j)}$ and $T_{mod(j)}$ denote observed and expected abundance of rank j, respectively. The model giving the best fit to each set of observed abundances was then determined as the model with the lowest value of D. Out of the 2000 samples, the relative gain of MACS relative to SRS is computed by:

$$g = 2000^{-1} \frac{I(D_{true} < D_{false} \mid MACS) - I(D_{true} - D_{false} \mid SRS)}{I(D_{true} - D_{false} \mid SRS)} \tag{6}$$

where $I(D_{true} < D_{false} \mid MACS)$ is the number of times the true model is chosen using MACS. Note that, for the purposes of this study, s is supposed to be known and empty samples (those that fall outside any clumps during the sampling) were discarded.

5 Results and Discussion

Results of the simulations performed are shown in Table 1 and 2 for the Geometric and the Brocken stick populations, respectively. In the tables are reported the expected sample size of MACS that is the sample size used for the SRS and the relative gain g as defined in (6). Perfomance of adaptive cluster sampling relative to simple random sampling was highest:

- in order to detect the Geometric model (Dominance of one species to the others) instead of the Brocken stick (Equitability);
- when the average final sample size was close to the initial sample.

For the less patchy population with many large clusters, i.e. $r = 4$, $T = 2000$ and $p = 5, 15$, the adaptive selection process resulted in a large percentage of the study area being sampled. On the other hand, for the most patchy populations, i.e. $r = 1, 2$, $p = 5$ and $T = 100, 500$, few quadrats were selected adaptively and the average final sample size was close to the initial sample. The Geometric model has more relatively rare species than the Brocken stick so that the sampling fraction is smaller indicating that adaptive sampling may be more favorable in low density population.

Furthermore, in the way we have chosen the aggregative condition, for the Brocken stick populations, the output is strongly dependent on the ability of detection clusters of different species whereas if the sampling effort would be determined by only one species, MACS performs very poorly in detecting equitability.

From a planning perspective the uncertainty of the final sampling fraction is a disadvantage of adaptive sampling designs. A solution might be the restricted adaptive sampling (Brown and Manly (1998)) where a limit is placed on the sample size prior

Table 1. Efficiency g of MACS with an initial sample of $n = 15$ relative to SRS with sample size $E(v)$ for the Geometric model. The population parameter were: T, the population total, p, the number of cluster, r, the average distance of an individual from a cluster center

T	p	r					
		1		2		4	
100	5	18	0.35	18	0.33	18	0.11
	15	18	0.09	18	0.15	18	0.21
500	5	37	0.40	60	0.25	55	0.17
	15	36	0.23	42	0.06	46	0
2000	5	92	0.25	195	−0.05	316	−0.19
	15	138	−0.01	182	−0.08	395	−0.14

Table 2. Efficiency g of MACS with an initial sample of $n = 15$ relative to SRS with sample size $E(v)$ for the Brocken stick model. The population parameter were: T, the population total, p, the number of cluster, r, the average distance of an individual from a cluster center

T	p	r					
		1		2		4	
100	5	18	0.14	18	0.03	18	0.11
	15	18	0.15	18	0.03	18	0.21
500	5	37	0	39	−0.01	41	0.17
	15	40	−0.04	42	−0.07	44	0
2000	5	112	−0.07	225	−0.15	319	−0.19
	15	175	−0.14	292	−0.11	343	−0.10

to sampling. Further improvements can be made adjusting the aggregative condition in order to consider the number of species detected (e.g. weighted sum of counts for several species). Finally, as pointed out by Thomposon (1992) pp. 275-276, for adaptive cluster sampling, comparisons should be made on the basis of cost because it is often less expensive to sample within a cluster than to sample a new cluster. Therefore, comparisons based on cost equations would tend to favor MACS more than comparisons based only on expected sample size as was done in this paper.

References

ANDREW, N.L. and MAPSTONE, B. D. (1987): Sampling and the description of spatial pattern in marine ecology. *Oceanographic and Marine Biology: An Annual Review*, *25*, 39–90.

BARABESI, L. and FATTORINI, L. (1998): The use of replicated plot, line and point sampling for estimating species abundances and ecological diversity. *Environmental and Ecological Statistics*, *5*, 353–370.

BROWN, J.A. and MANLY, B.J.F. (1998): Restricted adaptive cluster sampling. *Environmental and Ecological Statistics*, *5*, 49–63.

COCHRAN (1977): *Sampling Techniques*. Wiley, New York.

DI BATTISTA, T. (2002): Diversity index estimation by adaptive sampling. *Environmetrics*, *13*, 209–214.

DIGGLE, P.J. and BESAG, J.E. and GLEAVES, J.T. (1976): Statistical analysis of spatial point patterns bymeans of distance methods. *Biometrics*, *32*, 659–667.

GOVE, J.H. and PATIL, G.P. and SWINDEL, B.F. and TAILLIE, C. (1994): Ecological diversity and forest management. In: G.P. Patil and C.R. Rao (eds): . *Handbook of Statistics*, *12*. Elsevier Science B.V., Amsterdam. pp. 409–462.

MAC ARTHUR, R.H. (1957): On The relative abundance of bird species. *Proceedings of the National Academy of Science*, U.S.A., *43*, 293–295.

MAY, R.M. (1975): Patterns of species abundance and diversity. In: M.L. Cody and J.M. Diamond (Eds): *Ecology and Evolution of Communities*, Cambridge.

PATIL, G.P. and TAILLIE, C. (1982): Diversity as a concept and its measurements. *Journal of The American Statistical Association*, *77*, 548–567.

PIELOU, E.C. (1975): *Ecological Diversity*. Wiley, New York.

PRESTON, F.W. (1948): The commonness and rarity of species. *Ecology*, *29*, 254–283.

SEBER, G.A.F. and THOMPSON, S.K. (1993): Environmental Adaptive Sampling. In: G.P. Patil and C.R. Rao (eds): . *Handbook of Statistics*, *12*. *Elsevier Science B.V., Amsterdam. pp. 201–220.*

THOMPSON, S.K. (1996): Adaptive Sampling. *Wiley, New York.*

THOMPSON, S. K. (1993): Multivariate aspects of adaptive cluster sampling. In G.P. Patil and C. R. Rao (Eds): Multivariate Environmental Statistics..

WILSON, J.B. (1993): Would we recognize a broken-stick community if we found one . Oikos, 67, *181–183.*

Adjusted Least Square Estimation for Noisy Images

Ippoliti Luigi[1] and Romagnoli Luca[2]

[1] Dipartimento di Metodi Quantitativi e Teoria Economica, Facoltà di Scienze Manageriali dell'Università "G. d'Annunzio" di Chieti-Pescara, Viale Pindaro, 65127 Pescara, Italy

[2] Istituto Nazionale di Statistica, Ufficio regionale per il Molise, Via Mazzini 129, 86100 Campobasso, Italy

Abstract. In this paper we consider the problem of parametric estimation of a linear model corrupted by measurement error. In order to take into account the biasing effects caused by the presence of an external source of error, we propose an *adjustment* to the least square estimator. The statistical properties of the adjusted estimator are then experimentally verified with respect to two models characterising the literature of images analysis.

1 Introduction

In this paper we consider the problem of signal extraction. The object is to infer the value of a signal from a record which is affected by superimposed noise. Before embarking on the main topics, we need to establish some basic results concerning minimum-mean-square-error prediction. The criterion which is commonly used in judging the performance of an estimator or predictor $\widehat{Y}$ of a random vector Y is its mean-square error defined by $E(Y-\widehat{Y})^2$. Let $\{Y_t\}$ be a sequence of random variables and use a linear function of the values in the information set $\mathcal{I}_t = \{X_{j/t}; j \in \delta_t\}$ to predict the value of $\{Y_t\}$. Then the prediction may be denoted by the linear projection of $\{Y_t\}$ on $X_{j/t}$:

$$\widehat{Y}_t = X_{j/t}\beta \tag{1}$$

The principle of orthogonality implies that the minimum-mean-square-error estimate of Y_t is obtained by finding the coefficients $\{\beta_j\}$ which satisfy the conditions

$$0 = E\{X_j^T(Y-\widehat{Y})\} = E(X_j^T Y) - E(X_j^T X_j)\beta \tag{2}$$

where X_j is the whole information set $\mathcal{I}=\{X_j; j \in \delta_t, t = 1, 2, ..., n\}$. The precise nature of the solution depends upon the extent of the information set which is determined by the index set δ_t; in practice, the set will comprise only a finite (and small) number of elements.

The paper is organised as follows: theoretical results are established in Section 2; though signal extraction is a very general issue, we restrict our attention to the field of spatial analysis. In Subsections 2.1 and 2.2, we shall deal with two different spatial models, and propose two least square estimators "adjusted" to take into account the presence of an external source of error. Finally, Section 3 closes the paper with some experimental results both from simulations and from real observed data.

2 Signal Extraction from a Finite Sample

Consider the case of a signal sequence Y_t and imagine that the observations Z_t are contaminated by a white-noise error η_t, which is assumed to be statistically independent of the Y_t. Then, for a set of n observations, we should have

$$Z = Y + \eta \tag{3}$$

If the sequence Y_t underlying the observations is serially correlated, then there will be some scope for deriving better estimates of its values than those which are provided by the sequence Z_t of the observations. Note that this signal extraction problem may concern classical time-series analysis as well as spatial and image analysis. In this paper, Y_{ij} represents the grey level at pixel (i, j) of a $N \times M(= n)$ image and we consider parameter estimation for two particular Markov Random Fields observed with additive independent identically distributed noise.

2.1 Gaussian Markov Random Fields

Gaussian Markov Random Fields (GMRFs) are very commonly used for modelling textures in image analysis and more generally in the field of spatial data analysis; for an introduction see

Besag (1974). We consider parameter estimation in the presence of independent identically distributed (i.i.d.) noise. For particular boundary conditions, the implementation of optimal parameter estimation techniques such as maximum likelihood (ML) estimation can be computationally expensive and problematic. Maximum pseudo-likelihood (MPL) estimation (Besag, 1977) provides a quick and often reasonably efficient method of parameter estimation. Here, we consider adjusted MPL estimation of GMRF parameters in the presence of noise, making an adjustment which takes into account the noise. The gray levels are distributed according to a zero mean GMRF if the distribution of Y is multivariate normal with conditional mean and variances

$$E(y_{ij}|y_{rs}:(r,s)\in\delta_{ij}) = \sum_{(r,s)\neq(i,j)} \beta_{rs}y_{rs};$$
$$Var(y_{ij}|y_{rs}:(r,s)\in\delta_{ij}) = \tau^2;$$

where δ_{ij} is the set of neighbours of pixel (i,j) (not including (i,j)). For a homogenous model (spatially invariant) $\beta_{rs}=\beta_{-r,-s}$ and $\beta_{00}=0$. The primary objective is to estimate the parameters of the $(q+1)$-vector θ containing the distinct β_{rs} parameters and τ^2 in the presence of a zero mean independent identically distributed noise with variance σ_η. A very quick estimation method is pseudo-likelihood estimation (Besag, 1977). The procedure accepts a trade-off between efficiency and simplicity. The pseudo-likelihood is the product of the conditional density functions of each pixel given the rest and it is well known that, in the Gaussian case, MPL estimates can be obtained from the solution of equation (2). So, given a realization z of a noisy GMRF, we minimize the following function

$$F(z;\beta) = \sum_{i=1}^{N}\sum_{j=1}^{M}(z_{ij}-X_{ij}\beta)^2 \tag{4}$$

where β is the $(q\times 1)$ spatial parameter vector and $\mathcal{I}_{\rangle|} = \{X_{rs/ij};$ $(r,s)\in\delta_{ij},\ i=1,...,N; j=1,...,M\}$, which we shall simply

define as X_{ij}, is the $(1 \times q)$-vector of observations on the neighbourhood of z_{ij}. For example, considering a second order GMRF, we have $\beta=(\beta_{01}, \beta_{10}, \beta_{11}, \beta_{1-1})^T$ and
$X_{ij}=\{(z_{i,j-1}+z_{i,j+1}), (z_{i-1,j}+z_{i+1,j}), (z_{i-1,j-1}+z_{i+1,j+1}), (z_{i-1,j+1}+z_{i+1,j-1})\}$. Returning to the general order GMRF, after differentiating (4) with respect to β, we exploit the statistical properties of the measurement noise η_{ij} to give the *Adjusted Least Square Estimation (ALSE)*:

$$\widehat{\beta} = \{\sum_{i=1}^{N}\sum_{j=1}^{M} X_{ij}^T X_{ij} - 2n\sigma_\eta^2 I\}^{-1}\{\sum_{i=1}^{N}\sum_{j=1}^{M} X_{ij}^T z_{ij}\} \tag{5}$$

where I is the $(q \times q)$ identity matrix. The adjusted conditional variance estimate is then given by

$$\widehat{\tau^2} = \frac{1}{n}\sum_{i=1}^{N}\sum_{j=1}^{M}(z_{ij} - X_{ij}\widehat{\beta})^2 - \sigma_\eta^2(1 + 2\widehat{\beta}^T\widehat{\beta}) \tag{6}$$

Note that $\sigma_\eta^2(1+2\widehat{\beta}^T\widehat{\beta})$ is the adjustment which takes care of the noise. The measurement noise variance σ_η^2 is an additional parameter which is to be estimated before applying *ALSE* procedure. Several methods have been proposed in literature; for a review see Olsen (1993). In (5) it is assumed that parameters lie inside the valid parameter region. However, as in many real textures, it happens that the high correlation in the field leads the parameters to lie on the boundary, defining an Intrinsic GMRF. The difficulties in parameter estimation may be overcome by considering constrained least squares estimation: following Künsch (1987), who considered the noise-free case, we propose a *Constrained ALSE*, which assumes the following expression:

$$\widetilde{\beta} = \widehat{\beta} - K^{-1}C^T(CK^{-1}C^T)^{-1}(C\widehat{\beta} - c) \tag{7}$$

where $K=\sum_{i=1}^{N}\sum_{j=1}^{M} X_{ij}^T X_{ij} - 2n\sigma_\eta^2 I$, and C is a $(p \times q)$-matrix which defines the linear system of the parameters which must be solved according to the constraints in the $(p \times 1)$ vector c. As a consequence, the constrained conditional variance estimate, $\widetilde{\tau}^2$, is identical to expression (6), with $\widetilde{\beta}$ replacing $\widehat{\beta}$.

2.2 Markov Mesh Fields

Markov Mesh (MM) fields (Kanal, 1980) are a subclass of Markov Random Fields (MRFs) that impose quarter plane causality constraints on the neighbourhoods. As an MRF, a Markov Mesh field may be defined through its conditional probability density function but with non-symmetric halfplane (NSHP) $\Theta^{ij}=\{(r,s):(r<i)\cup(r=i,s<j)\}$. More generally, one could select the neighbours set as a subset of the quarter plane, since there is no unique definition of causality in two dimensions. A Gaussian Markov Mesh may be defined directly as a one-sided white noise driven Autoregressive field. Here we give an example of the third order NSQP (Non-symmetric Quarterplane) MM that has appeared frequently in image processing work. This field is given by

$$Y_{ij}-(\gamma_{0,-1}Y_{i,j-1}+\gamma_{-1,0}Y_{i-1,j}+\gamma_{-1,-1}Y_{i-1,j-1})=\varepsilon_{ij} \quad (8)$$

with coefficients $\gamma_{0,-1}$, $\gamma_{-1,0}$ and $\gamma_{-1,-1}$ representing, respectively, the interaction with the horizontal, vertical and left diagonal neighbours, see Figure 1. The driving noise input variables ε_{ij} is an independent, zero mean, Gaussian with variance σ_ε^2 and statistically independent of Y_{ij}.

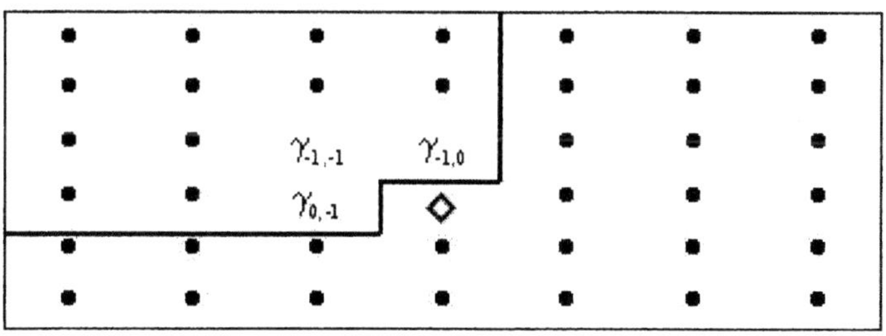

Fig. 1. Coefficients set for MM in (8).

Also in this case, the primary objective is to estimate the parameters in the presence of a noisy field. Considering model (8) and given a realization z of a noisy MM, it becomes natural to apply the *ALSE* as for GMRFs. The procedure, based on the new information set and Dirichlet boundary conditions (Balram and Moura, 1993) gives

$$\widehat{\gamma} = \{\sum_{i=1}^{N}\sum_{j=1}^{M} X_{ij}^T X_{ij} - n\sigma_\eta^2 I\}^{-1}\{\sum_{i=1}^{N}\sum_{j=1}^{M} X_{ij}^T z_{ij}\} \tag{9}$$

with an adjusted variance estimate

$$\widehat{\sigma}_\varepsilon^2 = \frac{1}{n}\sum_{i=1}^{N}\sum_{j=1}^{M}(z_{ij} - X_{ij}\widehat{\gamma})^2 - \sigma_\eta^2(1+\widehat{\gamma}^T\widehat{\gamma}) \tag{10}$$

3 Experimental Results

In this section we present parametric estimation results obtained by means of the Adjusted Least Square Estimator. Subsection 3.1 shows two simulation studies to asses the performance of the method when applied to the models presented above. Finally, in Subsection 3.2 two real data sets will be taken into account.

3.1 Simulation Results

Table 1 contains the results from parametric estimation on 100 simulated samples (lattice size 64×64) of second order homogenous GMRF, with parameters: $\beta_{0,1} = 0.05$, $\beta_{1,0} = 0.2$, $\beta_{1,1} = -0.15$, $\beta_{1,-1} = 0.05$, $\tau^2 = 100$. Gaussian zero mean i.i.d. noise was added, with variance =11.62, to produce an approximate SNR (Signal to Noise Ratio) of 10 dB. We also generated the same number of samples, on equal lattice size, for the MM model in (8). Results are reported in Table 2. The real parameters were fixed to: $\gamma_{0,-1} = 0.1$; $\gamma_{-1,0} = 0.5$; $\gamma_{-1,-1} = -0.1$; $\sigma_\epsilon^2 = 100$, with added Gaussian zero mean noise of variance $\sigma_\eta^2 = 12.19$. In each of the experimental trials, we fixed the added noise variance to its real value.

As it can be noted, ALSE appears to provide good results in both cases. The only point to note is related to a small underestimation of the driving noise variance for the MM model.

Table 1. Parametric estimation for second order GMRF.

GMRF Parameters	$\widehat{\beta_{0,1}}$	$\widehat{\beta_{1,0}}$	$\widehat{\beta_{1,1}}$	$\widehat{\beta_{1,-1}}$	$\widehat{\tau^2}$
Mean	0.048	0.200	-0.149	0.049	100.028
Standard Error	0.015	0.013	0.015	0.015	2.998

Table 2. Parametric estimation for MM (8).

GMRF Parameters	$\widehat{\gamma_{0,-1}}$	$\widehat{\gamma_{-1,-1}}$	$\widehat{\gamma_{-1,0}}$	$\widehat{\sigma_\epsilon^2}$
Mean	0.103	-0.100	0.496	96.130
Standard Error	0.017	0.019	0.014	2.987

3.2 Image Analysis Application

In this section two real images are considered to be realisations of noisy GMRFs. This assumption allows for the application of the adjusted estimator presented in Section 2.

The first image is a noisy Oak texture of size (128×128) that, in the framework of non-causal GMRF (Section 2.1), is employed to compare ALSE with respect to the corresponding Adjusted Maximum Likelihood Estimator (AMLE) presented in Dryden et al. (2001). In this case, the objective is to observe how "far" ALSE is from the best adjusted estimator AMLE. The noise-free image and the corresponding estimated parameters are presented in Figure 2 and Table 3 respectively.

Table 3. Parametric estimation for the Noise-free Oak texture.

First Order GMRF	$\widehat{\beta_{0,1}}$	$\widehat{\beta_{1,0}}$	$\widehat{\tau^2}$
Ordinary Least Square	-0.008	0.476	32.79
Maximum Likelihood	0.014	0.484	28.18

The well defined vertical structure of the texture has suggested the adoption of a first order GMRF. As it can be seen from Table 3, the ordinary least square estimates are very close to those of the maximum likelihood highlighting, as was expected, a strong vertical interaction parameter β_{10}. The noisy texture has been obtained by adding a zero mean Gaussian noise with vari-

Fig. 2. Noise-free Oak texture.

ance $\sigma_\eta^2 = 50$ obtaining a log signal-to-noise ratio (LSNR, Huang, Cressie; 2000) equal to 5. The results of the estimation procedure are shown in Table 4.

Table 4. Parametric estimation for the Noisy Oak texture.

First Order GMRF	$\widehat{\beta_{0,1}}$	$\widehat{\beta_{1,0}}$	$\widehat{\tau^2}$
ALSE	-0.018	0.481	31.73
AMLE	0.013	0.485	27.50

Considering the quite strong log signal-to-noise ratio, ALSE and AMLE perform very well, and in particular they continue to provide very similar results. Furthermore, taking into account that the computational complexity is $O(n)$ for ALSE and $O(n^2)$ for AMLE, we conclude that, particularly for large images, ALSE can be applied with a very quick algorithm at a low cost of precision. For a deeper analysis of the statistical properties of ALSE and AMLE the reader may refer to Dryden et al. (2001).

Due to its good features, ALSE has been integrated in a full image reconstruction procedure. In this case the (128×128) Lenna image (a classical benchmark) has been considered to compare a first order GMRF with a third order NSQP Gaussian Markov Mesh model. The noisy images are shown in Figure 3.a and 3.b.

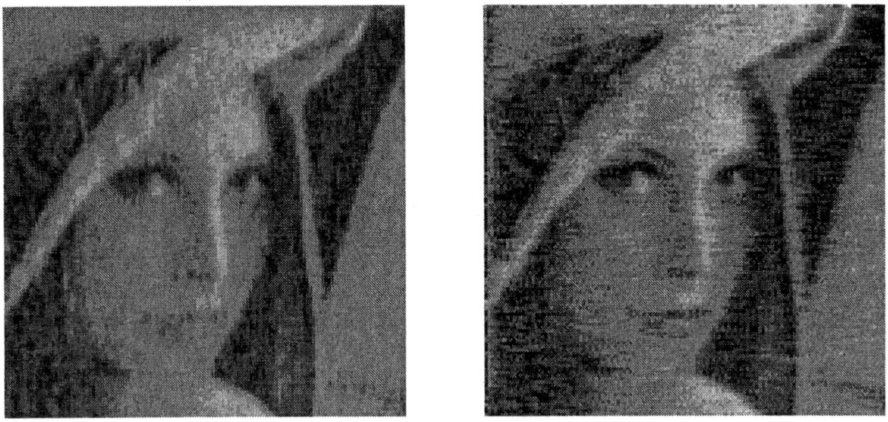

(a) (b)

Fig. 3. (a) Free-noise and (b) noisy Lenna

The noisy version is characterised by an added Gaussian noise with variance $\sigma_\eta^2 = 850$, producing an initial LSNR$\approx$ 5.5. Table 5 shows the results of the estimation procedure for the two models.

Table 5. Parametric estimation of noisy Lenna. Left: first order GMRF; right: third order NSQP GMM .

$\widehat{\beta_{0,1}}$	$\widehat{\beta_{1,0}}$	$\widehat{\tau^2}$		$\widehat{\gamma_{0,-1}}$	$\widehat{\gamma_{-1,0}}$	$\widehat{\gamma_{-1,-1}}$	$\widehat{\sigma_\epsilon^2}$
0.037	0.461	142.83		0.573	0.905	-0.487	126.39

The estimated parameters have then been used as input values of the state space model defined in Moura and Balram (1992). In this context, the Kalman filter proves to be a very efficient filtering and smoothing algorithm. The reconstructed images are presented in Figure 4.

The effectiveness of the reconstruction procedure seems evident for both models. On the other hand, it must be pointed out that the causality of the GMM involves an artificial directionality

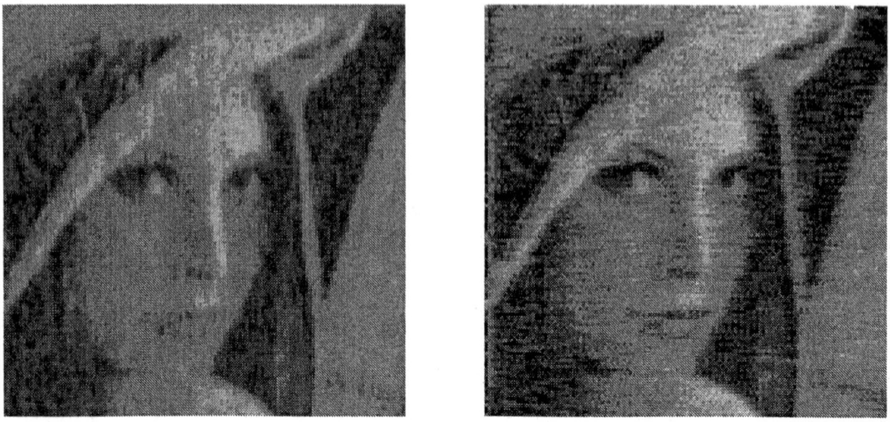

(a) (b)

Fig. 4. Reconstructed Lenna image: (a)GMRF (b) Gaussian MM

which is not present in the GMRF. This consideration, based on a pure visual analysis, is confirmed by a higher final LSNR for the GMRF which is equal to 9.32 versus 7.57 of the GMM.

References

BALRAM, N. and MOURA, J.M.F. (1993): Noncausal Gauss Markov Random Fields: Parameter Structure and Estimation. *IEEE Trans. Inform. Theory*, *39*, 1333–1355.

BESAG, J. (1974): Spatial Interaction and the Statistical Analysis of Lattice Systems. *J.R.S.S.*, *B*, *36*, 192–236.

BESAG, J. (1977): Efficiency of Pseudolikelihood Estimation for Simple Gaussian Fields. *Biometrika*, *64*, 616–618.

DRYDEN, I., IPPOLITI, L. and ROMAGNOLI, L. (2002): Adjusted Maximum Likelihood and Pseudo-Likelihood Estimation for Noisy Gaussian Markov Random Fields. *Journal of Computational and Graphical Statistics*, *11(2)*, 370–388.

HUANG, H.C. and CRESSIE, N. (2000): Deterministic/Stochastic Wavelet Decomposition for Recovery of Signal from Noisy Data. *Technometrics*, *42*, 262–276.

KANAL, L. (1980): Markov Mesh Models. *Comput. Graph. Image Process.*, *12*, 371–375.

KÜNSCH, H. (1987): Intrinsic Autoregressions and Related Models on the Two Dimensional Lattice. *Biometrika*, *74*, 517–524.

MOURA, J.M.F. and BALRAM, N. (1992): Recursive Structure of Noncausal Gauss-Markov Random Fields. *IEEE Trans. Inform. Theory*, *38*, 334–354.

OLSEN, S.I. (1993): Estimation of Noise in Images: an Evaluation. *Graphical Models and Image Processing*, *55(4)*, 319–323.

Flexible Dynamic Regression Models for Real-time Forecasting of Air Pollutant Concentration

Pietro Mantovan and Andrea Pastore

Dipartimento di Statistica - Università Ca' Foscari Venezia - S. Polo, 2375 - I-30125 Venezia (Italy)

Abstract. The application of the dynamic regression model to real-time forecasting of air pollutant concentration points out some problems due to both the high frequency of sampling and the need of many-step-ahead forecasting. Some flexible definitions of the system equation are proposed to solve these problems. The proposed definitions are evaluated by means of an application to the prediction of nitrogen dioxide concentration in Venezia-Mestre.

1 Introduction

Air pollutant concentrations and other related variables are usually collected in real time and with high frequency of sampling from monitoring network. One of the principal aims in this context is to produce real time forecasting of the pollutant concentrations, in order to evaluate the probability of crossing over warning or high risk thresholds. A statistical model cannot represent exactly the underlying chemical-phisical process. Therefore, flexible models or models with varying parameters can be defined. Dynamic regression models (Harrison and Stevens, 1976; West and Harrison, 1997) are a class of flexible models useful for this kind of problems. In fact, they allow to both model many kinds of dependence between observed variables (time dependencies, space dependencies, space-time dependencies) and include some singularities like outliers, influential data and change points. In these models, the system equation describes a law of evolution of the regression parameters vector over the time. In the paper, a more flexible definition of the system equation is proposed, which considers the *many-step-ahead* prediction errors within a sequential filtering algorithm. The proposed solutions are evaluated considering an application to the real time prediction of nitrogen dioxide concentration in Venezia-Mestre. Mestre is the dry-land part of Venice.

2 Dynamic Regression Models and Real Time Prediction

Consider a time series y_t, $t = 1, 2, \ldots$ and suppose to be mainly interested in the prediction, at each time t, of $y_{t+1}, y_{t+2}, \ldots, y_{t+s}$. To this purpose the values of a set x_t of k regressors are available, at each time t, and denote with θ_t the the k-dimensional vector of regression parameters at time t. The dynamic linear regression model is a generalisation of standard static regression model and it is defined by the following state-space representation:

$$y_t = x_t'\theta_t + e_t \tag{1}$$

$$\theta_t = M_t\theta_{t-1} + u_t \tag{2}$$

provided a vector of parameter θ_0 at an initial time $t = 0$ can be considered. e_t and u_t are random errors, usually assumed sequentially and mutually uncorrelated.

Equation (1) is the measurement (or observation) equation and is an usual regression equation, where the parameters vector θ_t is not constant, for all t. Following the terminology of the state-space models, θ_t is the state vector and e_t is the measurement error.

Equation (2) is the system (or transition) equation and describes a law of evolution of the regression parameters vector from one time to the following, where u_t is the system (or transition) error. The state-space representation of dynamic regression models is quite general and allows to consider general methods for inference (filtering) and prediction from a wide literature about dynamic system. A seminal method, the Kalman filter (Kalman, 1960; Harrison and Stevens, 1976), allows the bayesian analysis of the dynamic regression model when $e_t \sim \mathcal{N}[0, v_t]$, $u_t \sim \mathcal{N}[0, W_t]$, and $p(\theta_0) = \mathcal{N}[\theta_0|m_0, P_0]$, provided M_t, v_t, W_t, m_0, P_0 are known.

Denote with $Y_t^0 = (Y_{t-1}^0, y_t^0)$ the information about y_t available at time t, for $t = 1, 2, \ldots$, with $Y_0^0 = \emptyset$. Following a Bayesian interpretation, the Kalman filter is essentially a set of recursive equations that, considering the posterior distribution $p(\theta_{t-1}|Y_{t-1}^0) = \mathcal{N}[\theta_{t-1}|m_{t-1}, P_{t-1}]$ of the state vector at time $t-1$ and the observed value y_t^0, gives the parameters m_t and P_t of the posterior distribution $p(\theta_t|Y_t^0) = \mathcal{N}[\theta_t|m_t, P_t]$ of the state vector at time t and the predictive distribution $p(y_{t+1}, y_{t+2}, \ldots, y_{t+s}|Y_t^0)$ of future observation of the time series. Within the recursions of the Kalman filter, transition equation allows to define the prior distribution $p(\theta_t|Y_{t-1}^0)$ from the posterior distribution $p(\theta_{t-1}|Y_{t-1}^0)$. Many extensions of the Kalman filter have been proposed within the Bayesian approach (West and Harrison, 1997). Recently, sequential Monte Carlo filters have been proposed as interesting alternatives to the Kalman filter (see: Doucet et al., 2001 for a comprehensive review). Sequential Monte Carlo filters allow to consider, in a quite general way, non-linearity in the measurement and system equations and non-normality of the errors.

3 Sequential Filtering and Hyperparameter Estimation

Some aspects have to be focused at this point, if we are interested in real-time prediction of $y_{t+1}, y_{t+2}, \ldots, y_{t+s}$. The first aspect is the recursive nature of the Kalman filter. At time t, the distribution of the regression parameters vector θ_t depends, given the posterior distribution $p(\theta_{t-1}|Y_{t-1}^0)$ only on the data observed at time t and not on the complete sequence Y_t^0. This is absolutely relevant when data are collected with high frequency as in air pollution monitoring networks. In these cases, the dataset quickly increases and the use of prediction methods which have to consider at each time t the complete sequence Y_t^0 could be difficult. The second aspect to be focused concerns the so-called hyperparameters: M_t, v_t, W_t, m_0, P_0. These parameters are assumed as known in the Kalman filter but they are not known in the applications. The expected value m_0 and the variance P_0 are the parameters of the prior distribution of the state vector at $t = 0$ and they can be set subjectively by the researcher, given the initial state of information. However, their importance in the posterior distribution $p(\theta_t|Y_t^0)$ become small as t increases. For the system parameter M_t some recursive models have been proposed (see: Mantovan et al., 1999, 2000). In the following, we will assume that:

$$M_t = I, \forall t \quad (3)$$

For the variance v_t of the measurement error, if y_t is a univariate time series, a standard Bayesian sequential learning analysis is possible, via a straightforward extension of the Kalman filter. If y_t is a multivariate time series, some approximate bayesian solutions have been proposed (Mantovan and Pastore, 1999). As regards the estimation of the variance W_t of the system error, most of the literature is devoted to non-sequential procedures. There are two relevant approaches. One is concerned with maximum likelihood estimation, obtained by means of numerical methods, like scoring and E.M algorithm (Mc Whorter et al., 1976; Shumway and Stoffer, 1982; Watson and Engle, 1983; Ruud, 1991). More recently, within the bayesian approach, many authors proposed the use of MCMC methods (Carlin et al., 1992; Carter and Kohn, 1994; Frühwirth-Schnatter, 1994). Therefore, there are not solutions for the sequential learning of the variance W_t.

4 The System Equation as a Tool for the Model Flexibility

A more relevant question concerns the meaning of the system equation itself. In most of control theory and engineering applications, where a wide part of literature about dynamic system have been produced, the system equation is the representation of a known law of evolution of a system. In the context of dynamic regression models, the system equation represents a suitable law of evolution of the regression parameters vector, which is introduced to allow the models to be more flexible. Many alternatives are possible, as illustrated by Hastie and Tibshirani (1993) with the related inferential issues. Assume, as in equation (2), the evolution of the system parameter to be linear and markovian. If the assumption (3) about M_t holds, then the specification of the variance of the system error becomes relevant. A simple solution has been proposed by West and Harrison (1997). The variance matrix W_t is set in order to obtain:

$$\mathrm{var}(\theta_t|Y_{t-1}^0) = (1+\lambda)\cdot\mathrm{var}(\theta_{t-1}|Y_{t-1}^0) \tag{4}$$

with $\lambda > 0$, or, equivalently:

$$\mathrm{var}^{-1}(\theta_t|Y_{t-1}^0) = \delta\cdot(\mathrm{var}(\theta_{t-1}|Y_{t-1}^0))^{-1} \tag{5}$$

with $\delta = (1+\lambda)^{-1}$. Equations (4) and (5) highlight that the variance of the prior distribution, at time t, is greater than the variance of the posterior distribution at time $t-1$ and that the ratio between these two variances may be specified (or *controlled*) by a parameter, either λ or δ . West and Harrison call δ the *discount* parameter and suggest setting $0.80 \leq \delta \leq 1$. This approach allows to *control* the effect of the system error variance, but it has to be set by the researcher. A sequential learning on δ can be defined via a multiprocess model which introduces mixture approximation of the parameters vector distribution at each time t (Harrison and Stevens, 1976; West and Harrison, 1997). Considering the application of the dynamic regression model to the prediction of real data with high frequency of sampling (for instance: air pollution monitoring data), two other problems arise:

1. *Frequency of transition.* In the Kalman filter the parameters vector has a transition each time that a new value of y_t is observed. In other words, the frequency of sampling of y_t is also the frequency of the transition of the parameters vector. But in many automatic monitoring systems, the frequency of measurement (or sampling) is defined by measurement instrument characteristics or as function of the measures cost and it can be considered as *independent* from the frequency of transition of the parameters vector.
2. *Many-step-ahead prediction errors.* In the Kalman filter the posterior distribution of the parameters vector depends on the one-step-ahead prediction error. In many cases, with high frequency data, the researcher is interested in many-step-ahead

predictions. For instance, for hourly data, 24-hours-ahead (one-day-ahead) predictions could be interesting. If possible, the learning rule on the parameters vector should include also many-step-ahead prediction errors.

5 Flexible Definition of the System Equation

In order to avoid these problems, it is possible to propose more flexible definition of the system equation. Depending on specific application, some basic ideas can be combined in a specific solution. Here, some ideas and a specific solution are illustrated, which allows defining sequential filtering procedures more flexible than the usual solutions, by suitable modifications of the transition phase. Some basic ideas can be proposed about the transition frequency of the parameters vector.

Fixed lag transition. In a simple way, the transition could be defined at a fixed lag, that is every k times, $k \geq 1$ (for instance, for hourly data, with $k = 12$ or $k = 24$) or according to fixed interval of times, considering some characteristics of the variable y_t. For instance, in a model for prediction of NO_2 hourly concentration the transition could be fixed in specific hours with respect to the vehicle traffic. The idea is quite simple, but, as detailed in the following, it could improve the *many-step-ahead* prediction.

Prediction-error-driven transition. The transition could happen only when the model has produced bad predictions, considering both one-step-ahead and *many-step-ahead* predictions. Let:

$$\hat{y}_{t|t-j} = \mathrm{E}\left[y_t | Y_{t-j}^0\right]$$

and

$$q_{t|t-j} = \mathrm{var}\left[y_t | Y_{t-j}^0\right]$$

be respectively the expected value and the variance of the predictive disribution of y_t at time $t-j$, $j > 0$. Moreover, let r_t be the number of observation at time t after the last transition. At time t, if $j \leq r_t$, $\hat{y}_{t|t-j}$ is the last *j-step-ahead* prediction provided by the model. A function:

$$h_t\left(y_t, \left\{\hat{y}_{t|t-j}, q_{t|t-j}\right\}_{j=1,\ldots,min(s,r_t)}\right)$$

of the prediction errors and a threshold τ could be defined, for a suitable value of s, such that at time t the transition happens only if $h_t(\cdot) \geq \tau$.

An example of definition of $h_t(\cdot)$ is the following:

$$h_t(\cdot) = \left[\sum_{j=1}^{\min(s,r_t)} \left(y_t^0 - \hat{y}_{t|t-j}\right)^2 \pi_{t,j} \right]^{1/2}, \qquad \text{with} \quad \sum_{j=1}^{\min(s,r_t)} \pi_{t,j} = 1 \tag{6}$$

Other definition of the function $h_t(\cdot)$ can include an evaluation of prediction errors based on some specified utility function. Some alternative solutions for $h_t(\cdot)$ can be found in Huang and Cressie (1996). This idea explicitly includes the *many-step-ahead* prediction errors in the filtering procedure.

Bayes-factor-driven transition. This approach was proposed as *automatic monitoring* by West (1986). Essentially, at each time, in the transition phase two model (with and without) are considered and are compared via the predictive Bayes factor, based on the one-step-ahead predictive distributions under the two models.

As to the variance of the system error, there are two solutions.

Discounting approach. The effect of the transition on the variance of the prior distribution $p(\theta_t|y_{t-1}^0)$ is *controlled* via the discount parameter (West and Harrison, 1997).

Optimal discount via the Bayes factor. An extension of the bayes-factor driven transition can be introduced. Assume that, at each time t, the effect of the transition can be represented by a subset λ_t of the parameters of the predictive distribution, that, for simplicity, can be denoted as $p_{\lambda_t}(y_t|Y_{t-1}^0) = p(y_t|Y_{t-1}^0, \lambda_t)$. Moreover, assume that if $\lambda_t = 0$ the transition has no effect. Then, the Bayes factor: $\eta_t(\lambda_t) = p_{\lambda_t}(y_t^0|Y_t^0)/p_0(y_t^0|Y_t^0)$ can be considered as a function of λ_t . At each time, an optimal value of λ_t may be defined as: $\hat{\lambda}_t = \arg\max_{\lambda_t} \eta_t(\lambda_t)$ under the constraint $0 < \lambda_t < \lambda^+$, to avoid the model to follows outliers. As for the *Bayes-factor-driven transition*, in many parametric cases, $\hat{\lambda}_t$ has a simple analytic form and the computation is very easy. These ideas can be combined as follows in a solution which both considers the *many-step-ahead* prediction errors and chooses an optimal effect of transition. At each time t, the transition is driven by a function $h_t(\cdot)$ of *many-step-ahead* prediction errors (as defined before), and by a suitable threshold τ for this function. If $h_t(\cdot) > \tau$, then the transition happens and an optimal value of λ_t can be evaluate defining a Bayes factor as follows:

$$\eta_t\left(\lambda_t, \min(s, r_t)\right) = \frac{p_{\lambda_t}\left(y_t^0, y_{t-1}^0, \ldots, y_{t-\min(s,r)}^0 | Y_{t-\min(s,r)}^0\right)}{p_0\left(y_t^0, y_{t-1}^0, \ldots, y_{t-\min(s,r)}^0 | Y_{t-\min(s,r)}^0\right)} \quad (7)$$

and getting: $\hat{\lambda}_t = \arg\max_{\lambda_t} \eta_t(\lambda_t, \min(s, r_t))$. Following this solution, the *decision* about the transition is a function of prediction errors, while the *amount* of the transition (in terms of variance increasing) is determined via the predictive Bayes factor computed for all the predictive distributions of the *current* model.

6 On Line Forecasting of Nitrogen Dioxide Concentrations

In order to illustrate the proposed solutions, we consider a set of data from the air pollution-monitoring network of Venezia Mestre. These data are collected in real time and in this context a statistical model is required to give *many-step-ahead* forecasts in a very short time. In fact, if the concentration of a pollutant becomes higher than a warning threshold, public authorities must implement public intervention (such that traffic and house heating restrictions) in order to reduce the concentration. The subset of data considered in this paper concerns hourly measures of NO_2 in one station of the network, in the period between 01.01.97 and 08.02.97 (936 observations). The following regressors were considered for the prediction of NO_2 at time t: a weekly periodic component estimated by traffic data, the thermal gradient at time $t-8$, the wind speed at time $t-1$. One-hour and 24-hours-ahead predictions (expected value of predictive distributions) have been obtained with different transition frequencies and with different solutions for the amount of the transition effect. The predictive

Table 1. Predictive performances of the fixed-lag transition model, for different values of k

Frequency of Transition	Fixed lag		
	$k = 1$	$k = 12$	$k = 24$
Effect of transition	Fixed $W = 0.1I$	Fixed $W = 0.1I$	Fixed $W = 0.1I$
no. of transition	936	78	39
24-hrs-ahead pred.	0.778	0.783	0.762
1-hr-ahead pred.	0.784	0.802	0.799

Table 2. Predictive performances of the prediction-error-driven transition model, for different weight functions

Frequency of Transition	Prediction errors	
	(a)	(b)
Effect of transition	Fixed $W = 0.1I$	Fixed $W = 0.1I$
no. of transition	64	119
24-hrs-ahead pred.	0.777	0.775
1-hr-ahead pred.	0.830	0.870

performance has been evaluated by means of the square correlation coefficient between observed and predicted value (both for 1-hour ahead and 24-hours ahead forecasting). The number of transition occurred have been also considered.

First, we considered the model with fixed-lag transition (every k times) and a fixed variance matrix $W_t = W = 0.1I$. Three values of k were considered: $k = 1$ (transition every hour - usual Kalman filter solution), $k = 12$ (transition at 7 a.m. and at 7 p.m. every day), and $k = 24$ (transition at 3 a.m., every day). The results, reported in Table 1, highlight as a different lag of transition may improve the predictive performance of the model. The lag of transition parameter could has the same importance and the same effect on the predictions than the system error variance.

Then, we considered a model with a fixed system matrix $W_t = W = 0.1I$, but with prediction error driven transition. We consider the function defined by equation (7) with two different choices for the weights $\pi_{t,j}$:

(a) at each time t, $\pi_{t,j}$ uniform, for all j;

Table 3. Predictive performances of the Bayes-factor driven transition model, for both fixed and variable effects of transition

Frequency of Transition	Bayes Factor		
	$k = 1$	$k = 12$	$k = 24$
Effect of transition	Fixed	Discount	Discount
	$W = 0.1I$	$\lambda = 0.05$	$\lambda = 0.25$
no. of transition	66	80	135
24-hrs-ahead pred.	0.761	0.780	0.782
1-hr-ahead pred.	0.826	0.813	0.833

Table 4. Predictive performances of the optimal discount transition model, for different weight functions

Frequency of Transition	Prediction errors	
	(a)	(b)
Effect of transition	Bayes Factor	Bayes Factor
no. of transition	67	106
24-hrs-ahead pred.	0.795	0.790
1-hr-ahead pred.	0.866	0.858

(b) at each time t:

$$\pi_{t,j} = \begin{cases} 1 \text{ if } j = \min(s, r_t) \\ 0 \quad \text{otherwise} \end{cases} \tag{8}$$

In the second case, we are particularly interested in the *24-hours-ahead* predictions. In both cases the threshold was set equal to 30. Results are Table 2. Here the predictive performances are better than the cases with fixed-lag transition. Note as the number of occurred transitions is not high, compared with the number of considered observations.

Models with *Bayes-factor driven transition* have been considered with both a fixed variance matrix $W_t = W = 0.1I$ and setting following the discount approach, with $\lambda = 0.05$ (corresponds to $\delta \approx 0.95$) and with $\lambda = 0.25$ (corresponds to $\delta \approx 0.80$). Results are reported in Table 3.

Finally, we considered the model with prediction error driven transition and optimal transition effect via the Bayes factor, again with the function defined by equation (7) and with the two different choices for the weights $\pi_{t,j}$ described before. The results, reported in Table 4, appears to be better than in the other cases, particularly for 24-hours-ahead predictions.

¿From these results it is quite clear that the predictive performances of the model can be improved with an accurate setting of both the frequency and the effect of the transition equation. The improvement is particularly noticeable for 24-hours-ahead prediction. Moreover, we notice that both prediction error and bayes factor driven transition models require a reduced number of transitions.

References

CARLIN, B., POLSON, N., and STOFFER, D. (1992): A Monte Carlo approach to nonnormal and nonlinear state-space modeling. *Journal of the American Statistical Association,* 87, 493–500.

CARTER, C.K. and KOHN, R. (1994): On Gibbs sampling for state-space models. *Biometrika,* 81, 541–553.

DOUCET, A., DE FREITAS, N., and GORDON, N. (Eds.) (2001): *Sequential Monte Carlo Methods in Practice.* Springer, New York.

FRÜHWIRTH-SCHNATTER, S. (1994): Data augmentation and dynamic linear models. *Journal of Time Series Analysis,* 15, 183–202

HARRISON, P.J. and STEVENS, C. (1976): Bayesian Forecasting (with discussion). *Journal of the Royal Statistical Society (Ser. B), 38,* 205–247.

HASTIE T. and TIBSHIRANI R. (1993): Varying coefficient model. *Journal of the Royal Statistical Society (Ser. B), 55,* 757–796.

HUANG, H-C and CRESSIE, N. (1996): Spatio temporal prediction of snow water equivalent using the Kalman filter. *Computational Statistics and Data Analysis, 22,* 159–175.

KALMAN, R.E. (1960): A new approach to linear filtering and prediction problems, *Journal of Basic Engineering, 82D,* 33–45.

MANTOVAN, P. and PASTORE, A. (1999): Marginal updating equation for measurement error variance matrix and state vector estimates in the dynamic linear model. In: H. Bacelar Nicolau et al., eds): *Applied Stochastic Models and Data Analysis Proceedings of the IX International Symposium,* Instituto Nacional de Estatstica, Portugal, 212–217.

MANTOVAN, P., PASTORE, A., and TONELLATO S. (2000): A comparison between parallel algorithms for system parameter identification in dynamic linear models. *Appl. Stochastic Models Bus. Ind, 15,* 369–378.

MC WHORTER, A., SPIVEY, W.A., and WROBLESKI, W.J. (1976): A sensitivity analysis of varying parameter econometric models. *International Statistical Review, 44,* 265–282.

RUUD, P.A. (1991): Extension of estimation methods using the EM algorithm. *Journal of Econometrics, 49,* 305–341.

SHUMWAY, R.H. and STOFFER, D.S. (1982): An approach to time series smoothing and forecasting using the EM-algorithm. *Journal of Time Series Analysis, 3*, 253–264.

WATSON, M.W. and ENGLE, R.F. (1983): Alternative algorithms for the estimation of dynamic factor, MIMIC and varying coefficient regression models. *Journal of Econometrics, 23*, 385–400.

WEST, M. (1986): Bayesian model monitoring. *Journal of the Royal Statistical Society (Ser. B), 48*, 70–78.

WEST, M. and HARRISON, P.J. (1997): *Bayesian Forecasting and Dynamic Models (2nd ed.).* Springer, New York

Author Index

P

R

S

V

Subject Index

G

H

I

K

L

M

N

P

R

S

T

V

W

GPSR Compliance

The European Union's (EU) General Product Safety Regulation (GPSR) is a set of rules that requires consumer products to be safe and our obligations to ensure this.

If you have any concerns about our products, you can contact us on ProductSafety@springernature.com

In case Publisher is established outside the EU, the EU authorized representative is:

Springer Nature Customer Service Center GmbH
Europaplatz 3
69115 Heidelberg, Germany

Batch number: 10704713

Printed by Printforce, the Netherlands